GUIDE TO THE
HUMAN BODY

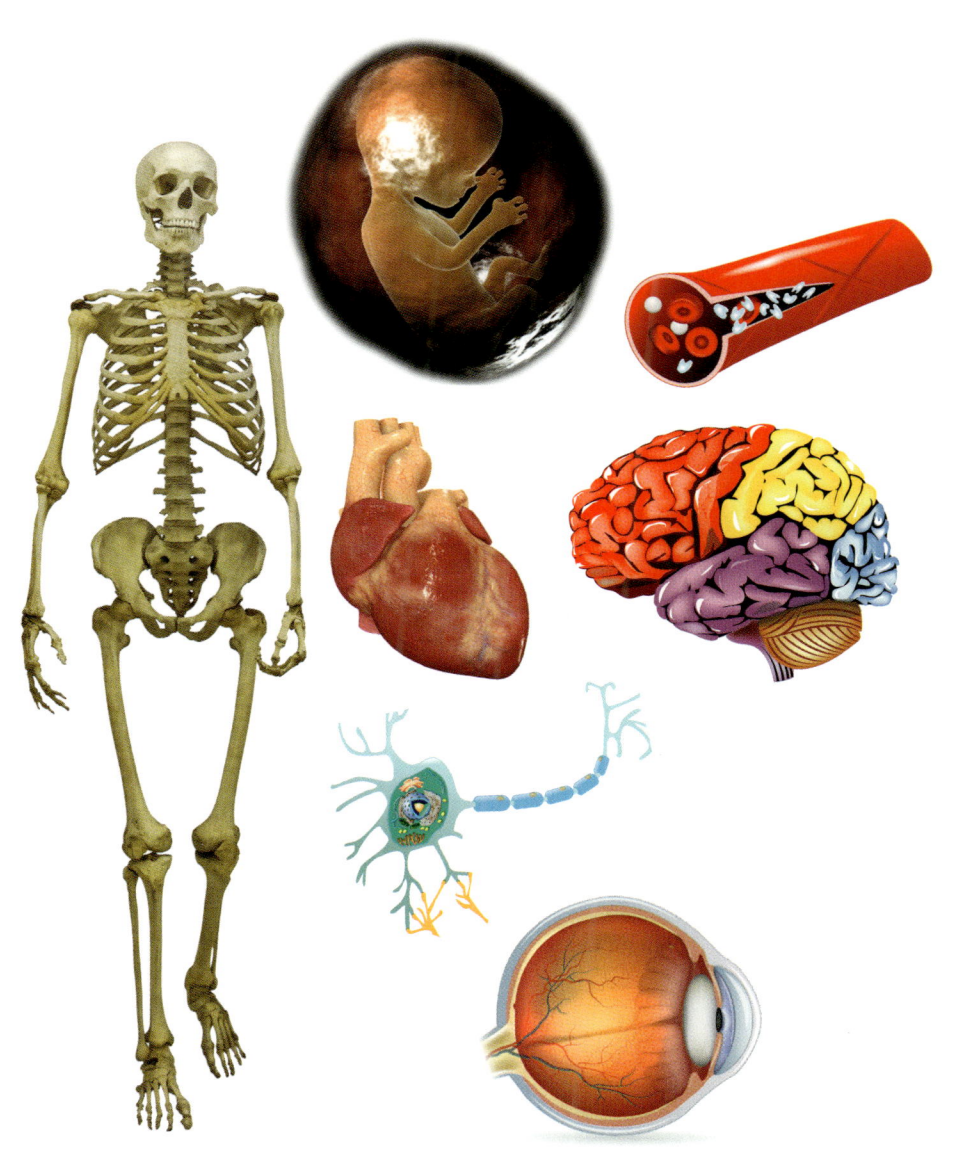

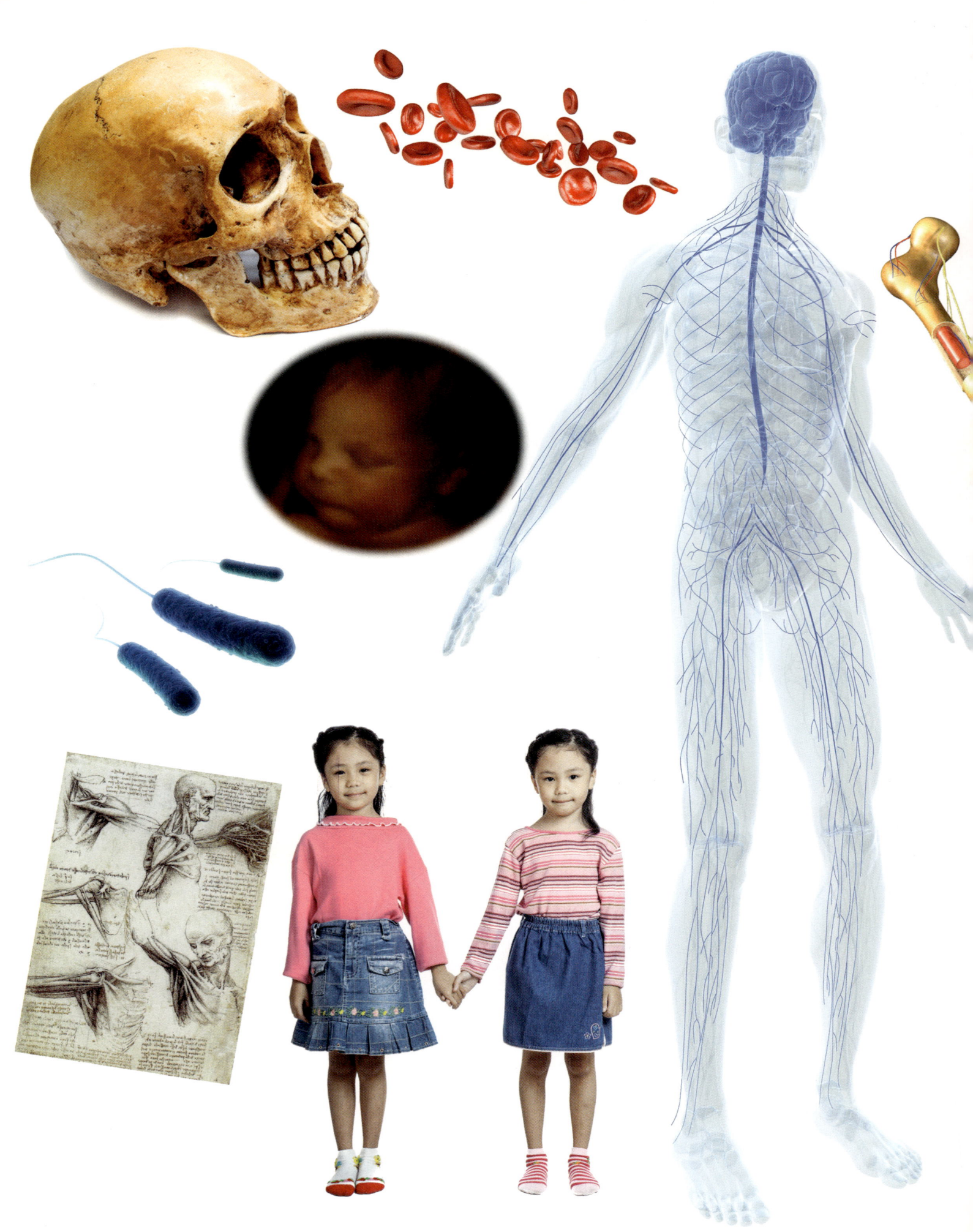

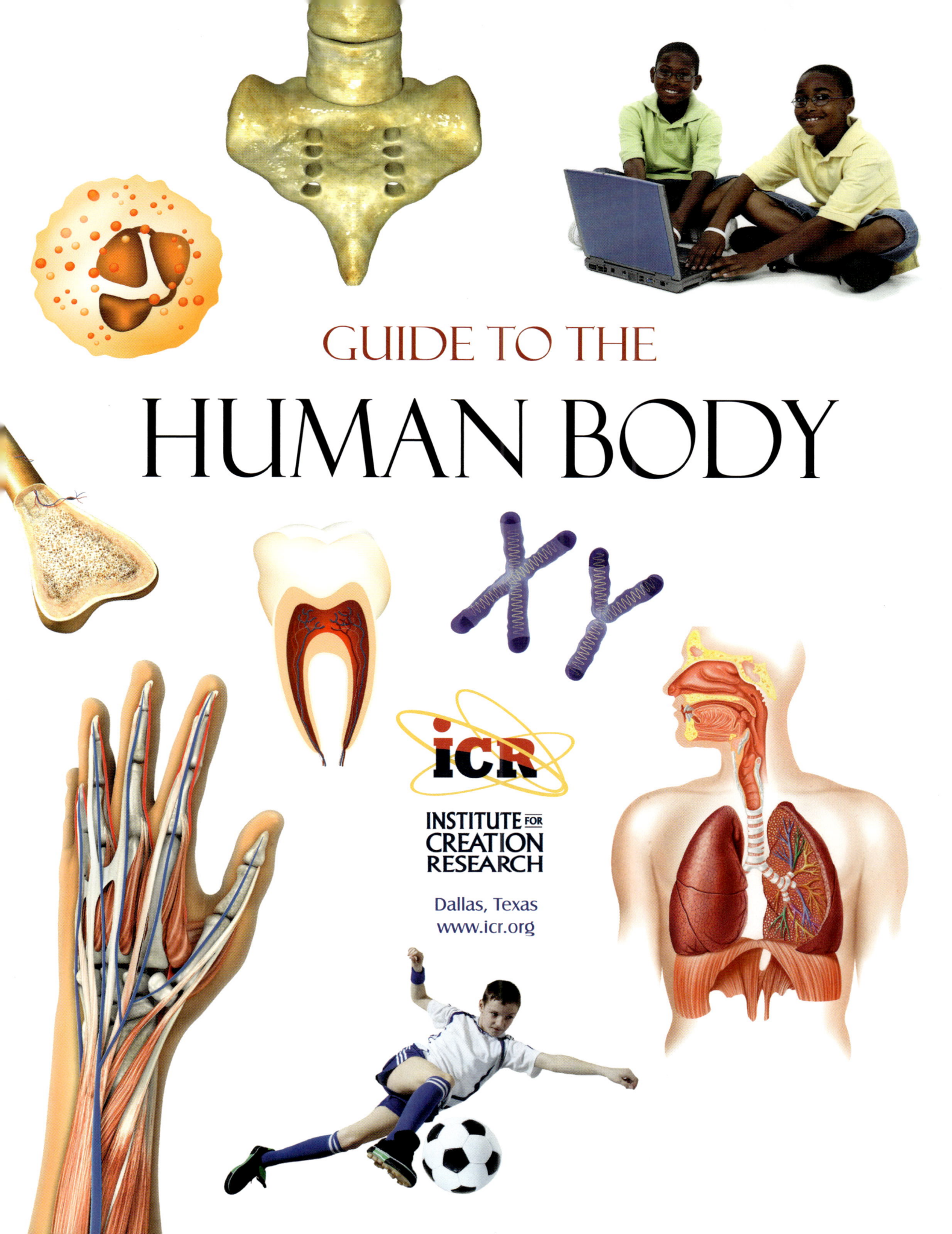

GUIDE TO THE
HUMAN BODY

icr
INSTITUTE FOR CREATION RESEARCH

Dallas, Texas
www.icr.org

Guide to the Human Body

First printing: November 2015

All Scripture quotations are from the New King James Version.

ISBN: 978-1-935587-80-4
Library of Congress Catalog Number: 2015950945

Please visit our website for other books and resources: www.icr.org

Printed in the United States of America.

Contents

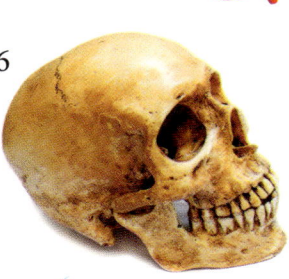

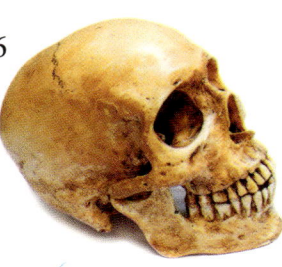

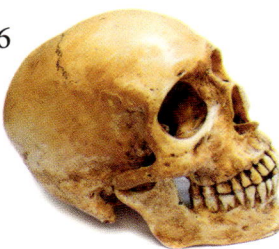

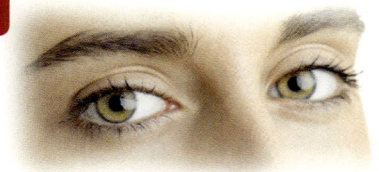

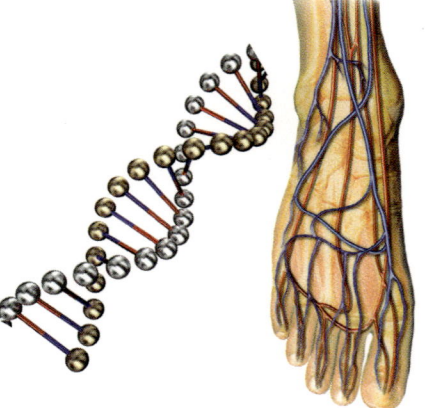

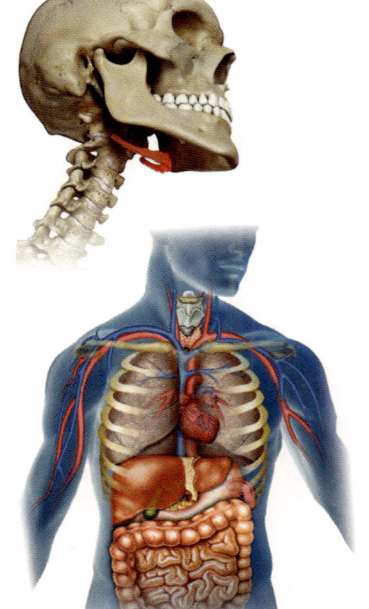
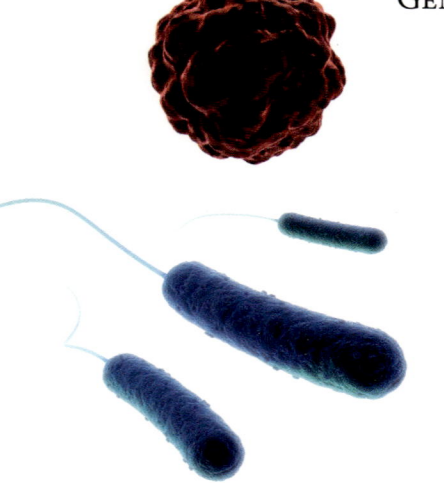

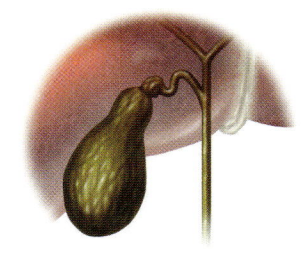

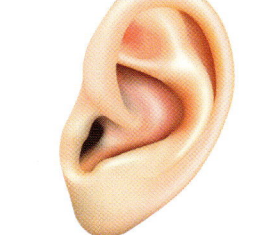

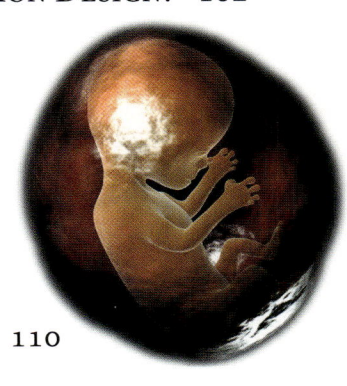

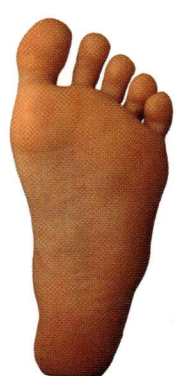

The Ultimate Designer: God

When we see anything that is complex in design and full of beauty—a painting, a sculpture, a purposeful machine with numerous interworking parts—what is usually the first question? In admiration and awe, we want to know, "Who made it?" Sweeping strokes of pink and blue framing a sunrise, blooming wildflowers on a green mountainside, and the peering blue eyes of a newborn baby cannot help but evoke this same question. Who made it? The Designer is not shy or reclusive, nor does He choose to remain anonymous. He lays claim to all these wondrous works in His Word, the Bible. The opening chapters tell us the universe and everything in it was designed in six days by an all-powerful God. His most prized creation is human beings, made fearfully and wonderfully in His image.

But you won't find this answer to "Who made it?" in most science books. Competing naturalistic theories attempt to challenge the source of everything we see. They do not credit the universe or the human body to a skillfull craftsman but instead claim that it all came into existence through a chaotic explosion. To some, the world is simply space rock that happens to swing around a burning ball of gas, and life is an accident, the result of chemical and biological mistakes. But that isn't what the evidence shows. Evolutionary ideas do not account for the intricate design of life, our abilities to appreciate beauty and to reason, or our need to find meaningful purpose in life.

"In the beginning God created the heavens and the earth." (Genesis 1:1)

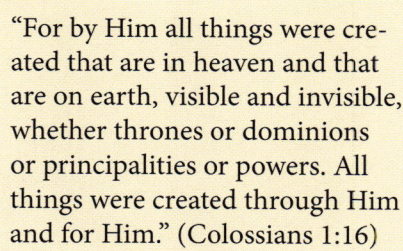

"For by Him all things were created that are in heaven and that are on earth, visible and invisible, whether thrones or dominions or principalities or powers. All things were created through Him and for Him." (Colossians 1:16)

Day 1   Day 2   Day 3

THE MYTH OF EVOLUTION

For a long time in the West, the literal six-day biblical creation was generally accepted, until 18th-century geologists began to question Noah's Flood. Later, Charles Darwin published *On the Origins of Species*, the book that shook the foundations of modern science. Darwin proposed that living creatures were not specially created but were instead the result of natural forces. His theory rests upon three main assertions. The first is that all biological life came from a single creature called a *universal common ancestor*. This creature multiplied and gradually changed into the variety of modern species we see today through a mechanism called *natural selection*. Natural selection, Darwin's second assertion, says that creatures naturally change in small ways in response to environmental pressures or other natural forces, and the ones more fit for their environment survive while the others die. In this way, life gradually evolved from simple to complex—step by tiny step. All of this supposedly happened over *a vast amount of time*, which is the third assertion. Darwin illustrated his theory in the tree of life, with the common ancestor at its root and evolution of other creatures as the branches.

"So God created man in His own image; in the image of God He created him; male and female He created them." (Genesis 1:27)

Thankfully, the whole universe—including the human body—proclaims the existence of a Creator through its sophisticated design that refutes slow, evolutionary, step-by-step construction by blind natural forces. Believing that God created it all best explains the intricate mechanisms we find in nature. Studying it with eyes wide open will satisfy many questions over origins and confirm God's inerrant and inspired Word.

"For since the creation of the world His invisible attributes are clearly seen, being understood by the things that are made, even His eternal power and Godhead, so that they are without excuse." (Romans 1:20)

As we examine the design of the human body, the answer to "Who made it?" becomes abundantly clear, since His signature of complexity, precision, and all-or-nothing unity has been written on every part.

"Not only must we understand that the universe and all that is in it were *created*, but we must know that everything has been *designed* by the omnipotent and omniscient God and has a purpose for being."

—Dr. Henry M. Morris III, CEO of the Institute for Creation Research

DAY 4

DAY 5

DAY 6

Why Study the Human Body?

Our beliefs about human origins profoundly affect how we approach things. If we think mankind is merely a highly evolved animal, then we have little reason to value a human life, its purpose, or its capabilities. However, if we view people as God's special creations, set apart for His glory, then we can see many reasons why the study of the human body is a significant endeavor.

FOR GOOD STEWARDSHIP

After God created humans, He made us stewards of the earth and its resources and also passed on the immense responsibility to take care of it. The fact that God provided the earth for us to use doesn't mean we should be reckless and irresponsible. Rather, we need to exercise wisdom and prudence. The same is true for our own bodies. Studying the way we are made helps us to understand how our bodies work, enabling us to become better stewards in the care and use of what He has entrusted to us.

TO SATISFY OUR THIRST FOR KNOWLEDGE

The way our bodies work is fascinating. They are made of living tissues and yet hold all the working features that man-made machines currently do. In fact, many tools and machines we use were inspired by our bodies' capabilities. Whether we're only mildly curious or intensely interested, we can all benefit from knowing more about the bodies we've been given.

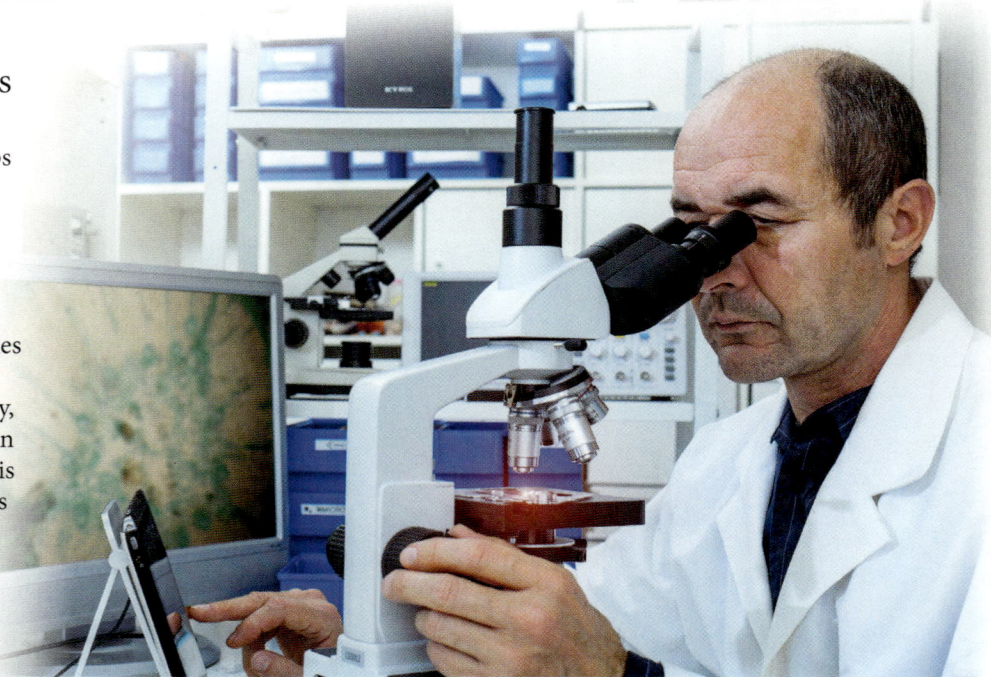

TO MAKE NEW DISCOVERIES

Our knowledge and understanding of the human body increased by leaps and bounds the past few centuries. Fervent study and research brought numerous breakthroughs—intricate workings of the body have been mapped, diseases once considered deadly are now treatable, and surgeries are less risky and invasive. And with the substantial increase in technology, we can view what is happening within the body from the outside. Yet there is still so much we don't understand. As people continue to study the human body, who knows what life-changing discoveries will be made?

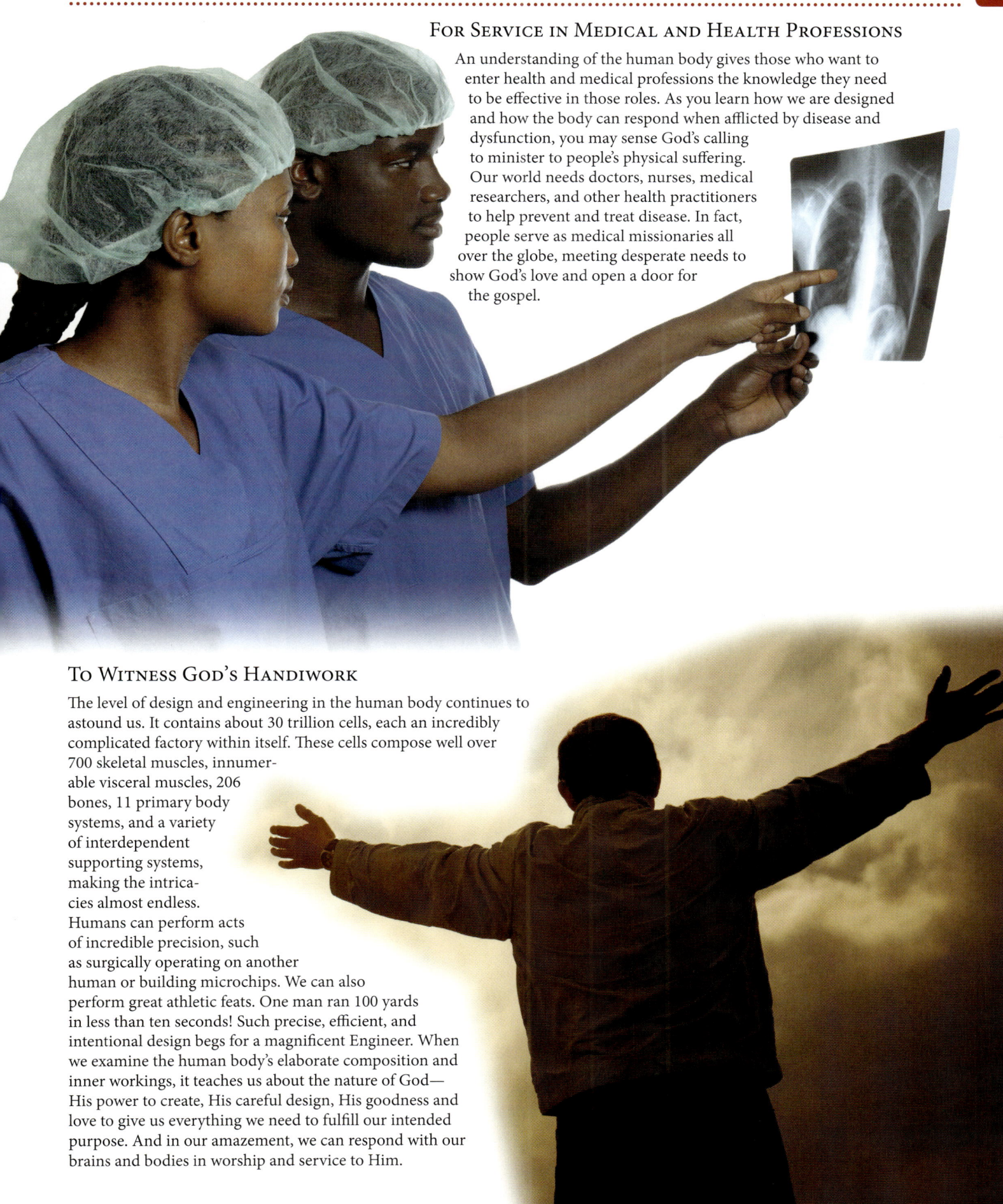

FOR SERVICE IN MEDICAL AND HEALTH PROFESSIONS

An understanding of the human body gives those who want to enter health and medical professions the knowledge they need to be effective in those roles. As you learn how we are designed and how the body can respond when afflicted by disease and dysfunction, you may sense God's calling to minister to people's physical suffering. Our world needs doctors, nurses, medical researchers, and other health practitioners to help prevent and treat disease. In fact, people serve as medical missionaries all over the globe, meeting desperate needs to show God's love and open a door for the gospel.

TO WITNESS GOD'S HANDIWORK

The level of design and engineering in the human body continues to astound us. It contains about 30 trillion cells, each an incredibly complicated factory within itself. These cells compose well over 700 skeletal muscles, innumerable visceral muscles, 206 bones, 11 primary body systems, and a variety of interdependent supporting systems, making the intricacies almost endless. Humans can perform acts of incredible precision, such as surgically operating on another human or building microchips. We can also perform great athletic feats. One man ran 100 yards in less than ten seconds! Such precise, efficient, and intentional design begs for a magnificent Engineer. When we examine the human body's elaborate composition and inner workings, it teaches us about the nature of God— His power to create, His careful design, His goodness and love to give us everything we need to fulfill our intended purpose. And in our amazement, we can respond with our brains and bodies in worship and service to Him.

History of Anatomy

Over thousands of years, people of many cultures and time periods have been intrigued by the human body and its design. Many beliefs, theories, and experiments have caused our understanding to change over time. Here are some famous characters in history who contributed to the greater understanding we have today.

HIPPOCRATES

In the 4th century B.C. a doctor lived in the Greek city of Cos who would later be called the Father of Medicine—Hippocrates. Born into a family of physicians, Hippocrates soon became proficient at the practice of medicine and was well-known throughout the Mediterranean area. He is credited with curing many diseases, including the tuberculosis that afflicted the king of Macedonia. Myth and truth are difficult to separate regarding his life, and many medical works may have been falsely attributed to him. However, it is known that he wrote around 60 documents on medicine collectively referred to as the *Hippocratic Corpus.* His life and works became the foundation of Western medicine and inspired today's Hippocratic Oath, in which beginning physicians swear to follow ethical standards in their practice of medicine.

GALEN

About 600 years after Hippocrates, another Greek physician revolutionized medicine. Galen was the chief physician to the gladiators in the city of Pergamum. This role gave him many opportunities to study the human body through the gruesome injuries of his patients. For further study, he dissected apes and pigs. Galen overturned the widely accepted idea that arteries carry air throughout the body, a belief based on the arteries of dead animals, which appeared empty. His expertise moved him to Rome, where his fame continued to rise through his public demonstrations on anatomy. He identified the valves of the human heart and differentiated arteries and veins. He successfully cured many people whom other doctors had given up on. The Roman joint emperors hired him as their physician for a military campaign in northern Italy.

Leonardo da Vinci

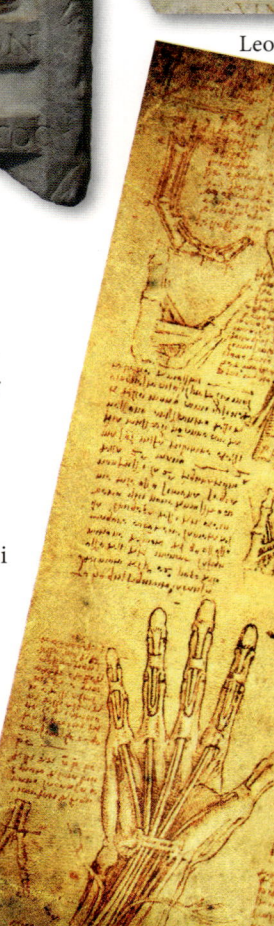

LEONARDO DA VINCI

After Galen, Western medical study was largely ignored—until the 15th century, when Leonardo da Vinci practically stumbled into the profession. He had built a successful career as an artist in Milan, Italy. A doctor named Marcantonio della Torre approached him to do the sketches for a book on human anatomy. The two collaborated, with della Torre dissecting human bodies and da Vinci sketching them. However, della Torre passed away unexpectedly before finishing the book, and so da Vinci took up both tasks. He dissected bodies in a cathedral cellar and sketched his findings. He created over 500 diagrams that still amaze physicians with their accuracy and depth of anatomical understanding.

Hippocrates

ANDREAS VESALIUS

The 16th century saw the rise of Andreas Vesalius, born in Belgium to a family of physicians and pharmacists. He studied in Paris at one of the leading medical schools. Soon afterward, he wrote seven works on human anatomy collectively called the *Fabrica*, lavishly illustrated and filled with cutting-edge research. It was primarily for this collection that he became known as the Founder of Modern Anatomy. After being shown the *Fabrica*, Emperor Charles V promoted Vesalius to his royal medical staff.

Andreas Vesalius

WILLIAM HARVEY

In the 17th century, Britain legalized dissections, and medical research began to flourish. One of the most prominent medical figures of this time was William Harvey, a physician who studied medicine in Italy. Harvey focused his research on the circulatory system and published a groundbreaking work titled *Anatomical Studies on the Motion of the Heart and Blood in Animals.* He argued against the archaic practice of blood-letting, and his career suffered from the backlash. Despite this, he eventually became known as London's best physician and was appointed to King James I's personal medical staff.

DID YOU KNOW?

Until the late 19th century, many physicians believed that withdrawing blood could cure or prevent illness. This may have contributed to George Washington's death.

William Harvey

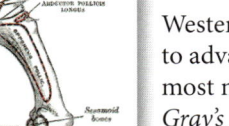

Henry Gray

HENRY GRAY

Western medicine continued to advance, with perhaps the most notable publication being *Gray's Anatomy* in 1858. Henry Gray worked with fellow physician Henry Vandyke Carter for 18 months in London on the book, using Vesalius' *Fabrica* as a model. It continues to be updated, published, and used as a medical reference on the human body. The year 2008 marked the 150th anniversary of the original *Gray's Anatomy* and the publication of its 40th edition.

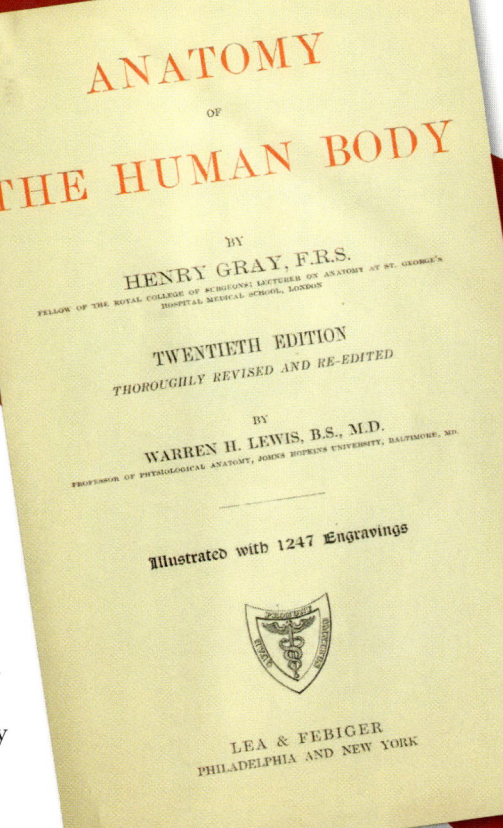

ANATOMY
OF
THE HUMAN BODY
BY
HENRY GRAY, F.R.S.
FELLOW OF THE ROYAL COLLEGE OF SURGEONS; LECTURER ON ANATOMY AT ST. GEORGE'S HOSPITAL MEDICAL SCHOOL, LONDON

TWENTIETH EDITION
THOROUGHLY REVISED AND RE-EDITED

BY
WARREN H. LEWIS, B.S., M.D.
PROFESSOR OF PHYSIOLOGICAL ANATOMY, JOHNS HOPKINS UNIVERSITY, BALTIMORE, MD.

Illustrated with 1247 Engravings

LEA & FEBIGER
PHILADELPHIA AND NEW YORK

The Big Picture: A System of Systems

If you see apples scattered on the ground under an apple tree, you might guess that they just happened to fall to their current locations. But if you see apples set on the ground with stems upright in a straight line, from smallest to largest, you would know that someone placed them there. Designed things can always be recognized by having multiple parts working together for a purpose. The more complex and well-ordered something is, the more it points to intentional design.

The human body is probably the single best example of complex and orderly design in the universe, since its parts were all purposefully orchestrated to function as a living whole. Your organs and their processes are not disorganized. They aren't scattered randomly throughout the body in no particular order. Quite the opposite is true.

Your body's amazing collection of 11 primary systems works seamlessly to sustain life and enables you to think, talk, feel, move, hear, see, eat, and breathe. Along with the primary systems, you possess supporting systems. This unified "system of systems" shows *all-or-nothing unity* in many ways. Also called *irreducible complexity,* all-or-nothing unity means that if a key part or process were missing, your body would not continue to function properly. Like apples in sequence on the ground, the intelligent arrangement of the body's systems tells us that someone intentionally placed them there.

The **respiratory** system brings oxygen into the body through the lungs and exhales carbon dioxide. Vital nutrients from food are absorbed, and waste is expelled through the **digestive** system.

The **circulatory** system constantly pumps blood carrying nutrients, oxygen, carbon dioxide, blood cells, and hormones throughout the body, supplying every cell with nourishment and removing waste. The **immune** system uses the lymphatic system to transport white blood cells and clear the body of toxins, waste, and leftover fluids. It is also supported by the **urinary** system, which removes these substances from the blood through the kidneys and carries them out of the body. The **endocrine** system's glands produce hormones to regulate processes, such as growth and development, sleep, reproduction, mood, and metabolism.

The **nervous** system connects the brain to the rest of the body and allows the mind to tell it when and how to move, bringing thought into action. Neural signals zip through the body's nervous system and give us the ability to react to our environment, while the **muscular** system enables us to run, write, play the piano, and hold hands.

Skin, the main part of our **integumentary** system, covers our entire body and acts as an interface with the outside world. The **skeletal** system supports our bodies with sturdy bones and joins to our muscular system with strong ligaments and tendons. The **reproductive** system enables us to produce new human beings made in the image of God.

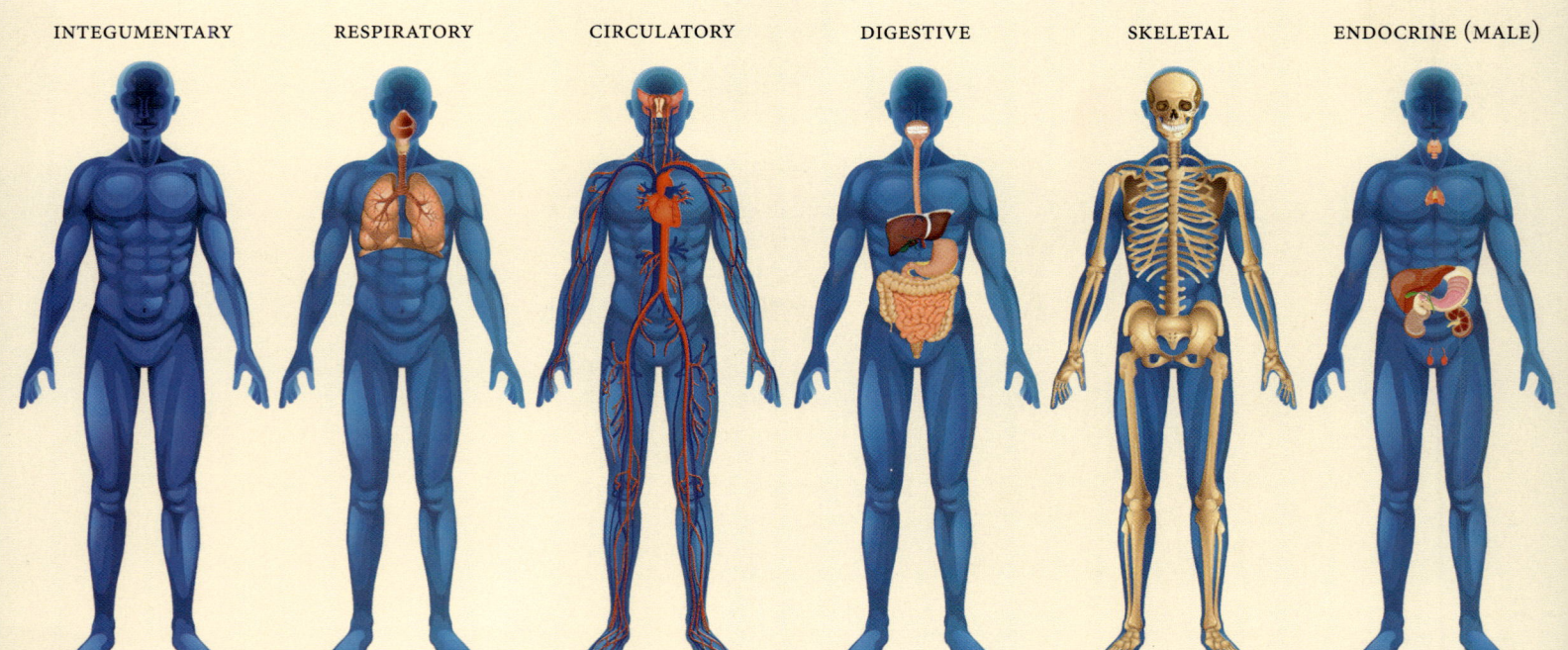

INTEGUMENTARY   RESPIRATORY   CIRCULATORY   DIGESTIVE   SKELETAL   ENDOCRINE (MALE)

DID YOU KNOW?

Renaissance artist Leonardo da Vinci (1452-1519) dissected human bodies to find out the mysterious ways they functioned.

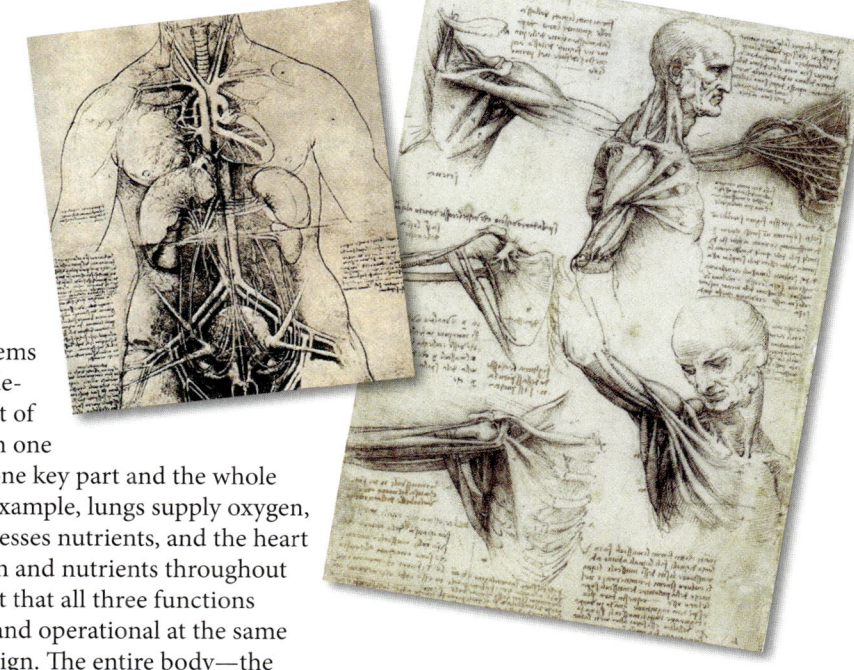

The human body's 11 systems are interdependent. Interdependent systems consist of parts that depend on one another. Remove one key part and the whole system fails. For example, lungs supply oxygen, the stomach processes nutrients, and the heart transports oxygen and nutrients throughout the body. The fact that all three functions must be present and operational at the same time points to design. The entire body—the system of systems—was initially created with all of its parts intact and fully functional at the moment of creation! Interdependence is exactly what we would expect to see if God purposely created living systems.

The theory of evolution says that natural processes built the human body systems bit by bit over millions of years. But evolution has no proven explanations for the origin of just one irreducibly complex system, let alone the interdependent web of 11 irreducible systems that comprise the human body. A wonderfully constructed human body is exactly what an all-wise Creator would make from the very beginning.

DID YOU KNOW?

As incredible as the human body is, God promised that those who trust in Him will one day inherit an even more incredible new body. "The Lord Jesus Christ… will transform our lowly body that it may be conformed to His glorious body" (Philippians 3:20-21).

"So God created man in His own image; in the image of God He created him; male and female He created them." (Genesis 1:27)

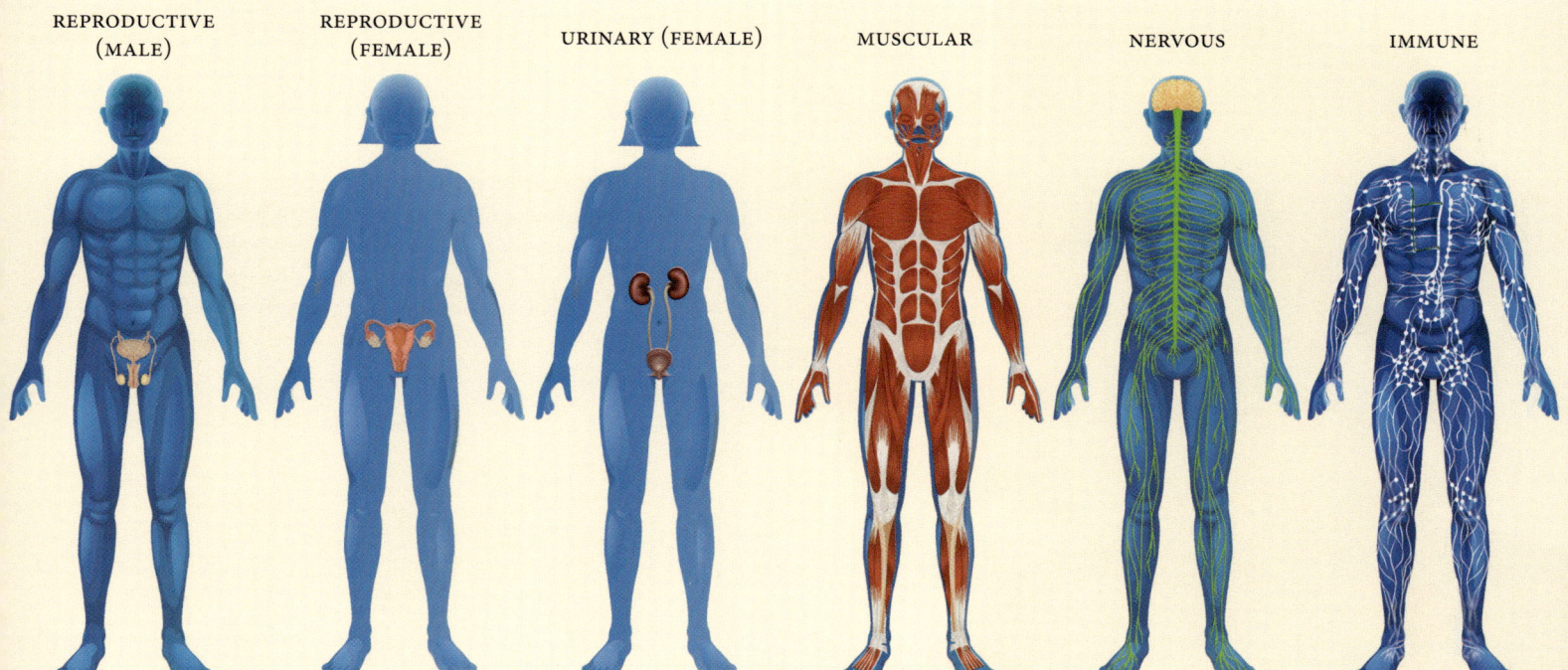

REPRODUCTIVE (MALE)   REPRODUCTIVE (FEMALE)   URINARY (FEMALE)   MUSCULAR   NERVOUS   IMMUNE

At First Glance

Although humans share a similar design on the inside, we display a vast array of differences on the outside. The next time you visit a state fair or crowded shopping mall, take a look around. From eye color to hair texture, freckles to fingerprints, the immense creativity of the Lord Jesus Christ can be seen in each one of us, even at first glance.

SKIN

Melanin is a pigment that affects the coloring of certain parts of the human body and is produced by cells called *melanocytes*. The more melanin you have, the darker your coloring will be. Likewise, paler complexions have less melanin. Eumelanin tends to color skin in shades of black and brown, while pheomelanin tends to make skin reddish or yellow.

EYES

The most common eye colors are green, brown, and blue, but occasionally people have eyes that are hazel, gray, or amethyst. Individuals with black or brown eyes have more melanin in their irises, which gives them greater protection from the sun. Those with lighter eyes have very little protection from melanin and may experience discomfort, irritation, burning, or even tissue damage if the eyes are not protected from the sunlight. Eye shapes vary as well—almond, down-turned, prominent, hooded, and wide-set shapes are just a few of the characteristics that can make eyes unique.

DID YOU KNOW?

Albinism is a genetic disorder in which a person is born with little to no melanin. They usually have pale skin, white hair, and pinkish eyes.

DID YOU KNOW?

The reason hair becomes gray as people get older is because it loses melanin.

HAIR

Melanin also determines hair color. Eumelanin tends to produce dark hair, and pheomelanin produces reddish hair. Having only a small amount of eumelanin produces blond or brown hair. The texture of hair—curly, wavy, or straight—is impacted by the shape of the hair follicle. Round follicles cause straight hair, oblong ones cause wavy hair, and skinny, oval-shaped follicles produce curly hair.

DID YOU KNOW?

Dimples are a dominant inherited trait. If your parents both have dimples, you probably will, too.

Did you know?

Fingerprints form in the womb when the dermal cell layer of your skin is squeezed between the inner subcutaneous tissue and the outer dermis. The pressure creates wrinkles within the dermal cell layer, which forms the loops, arches, and ridges that make your prints one of a kind.

Freckles

Some skin melanocytes produce more melanin than others, causing spots called freckles. Freckles become dark when exposed to the sun, but usually diminish over time without sun. People with red hair and green eyes tend to have greater amounts of pheomelanin so they are more prone to have freckles. Though most freckles are harmless, individuals should get regular checkups on any that appear to be changing or are multi-colored or bigger than a pencil eraser to guard against melanoma.

On top of the numerous traits seen in our skin, hair, and eyes, other features vary from person to person. The Lord Jesus Christ built the potential for diversity into Adam and Eve's genes, beautifully expressed in their descendants all around the world. Some may be surprised that such a wide range of traits could be passed down from only two people, but this is perfectly consistent with modern genetics. Many genes influence our physical traits, including at least three producing pigments that darken skin color. If Adam and Eve carried three genes for pigment and three genes for no pigment, then they would have had medium-brown skin—the most common tone still seen today. The chart below illustrates how different combinations of Adam and Eve's genes in this scenario could express all known skin tones in their children.

Did you know?

The integumentary system includes the skin, hair, and nails. The skin is the largest organ in the human body and consists of layers of tissue: the epidermis, dermis, and hypodermis.

"After he begot Seth, the days of Adam were eight hundred years; and he had sons and daughters." (Genesis 5:4)

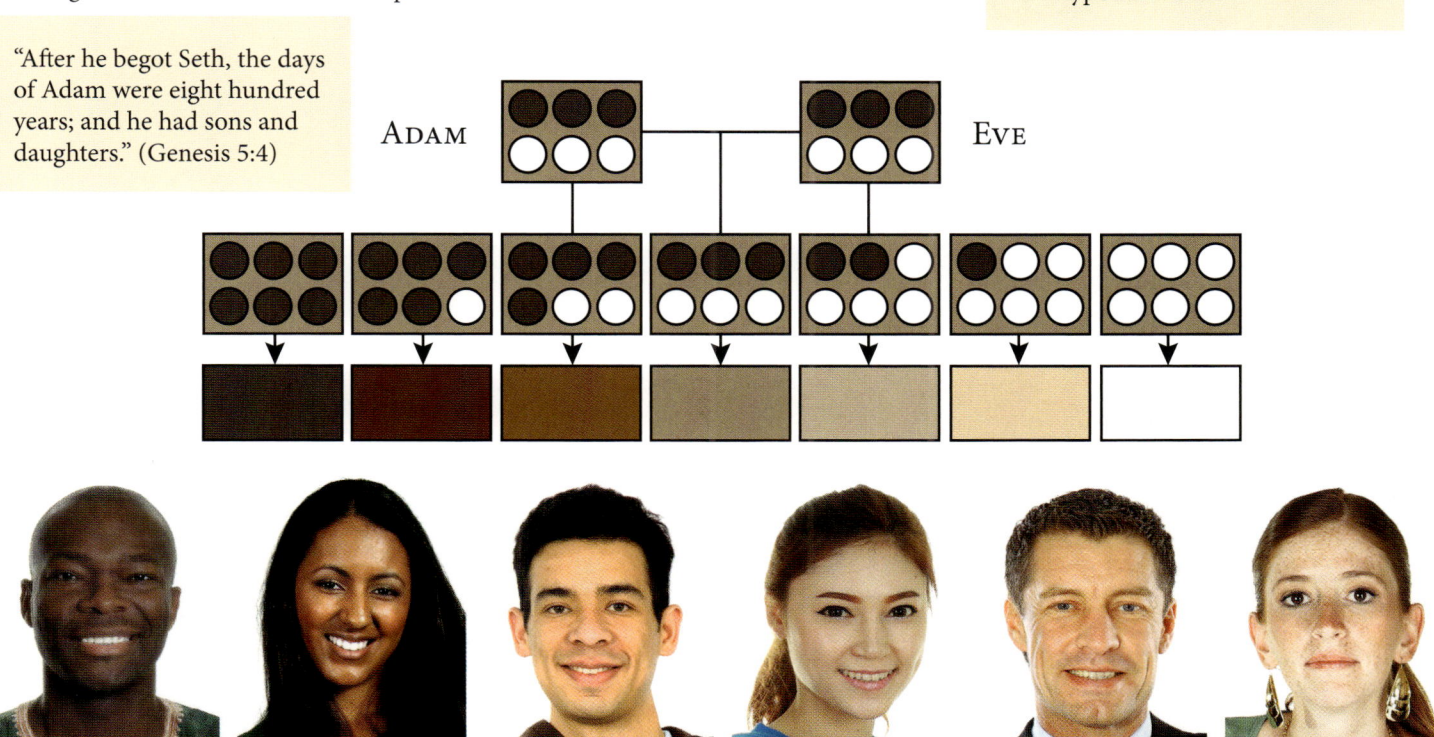

The Skeletal System: A Strong Support

A human body without its skeleton would be like a tent without the poles, a house without a frame, a glove without a hand, a dress for sale without a mannequin…well, you get the idea. Your skeleton—made up of bones—helps you move, protects your insides, and supports soft tissues like skin and muscles. Without it, you'd be a limp bag of organs. It also stores calcium, iron, and energy on the insides of its 206 individual bones, which are arranged into two main categories: the axial skeleton, which supports the head and torso, and the appendicular skeleton, which supports the arms and legs.

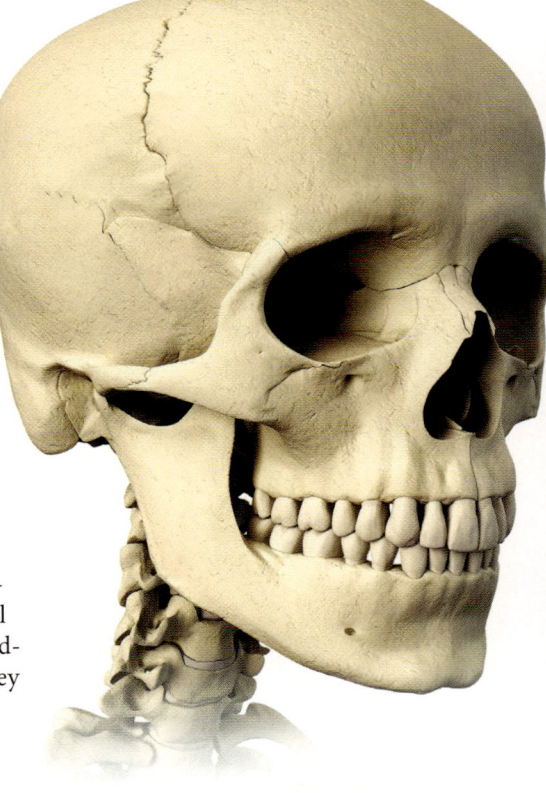

Skull

The skull consists of 22 bones, serves as a shield to protect your brain, and plays a significant role in the appearance of your facial features. The bones are separate during childhood while the head is still growing, but they slowly bind together at conjunctions called *sutures* as the body matures, providing added strength and support in adulthood.

Vertebrae

About 33 individual bones—referred to as vertebrae—make up the vertebral column lining your neck, back, and lower torso. If the column were composed of one long bone instead of small flexible vertebrae, you couldn't even bend over to put on your shoes. Categorized into sections, the top seven vertebrae near the neck are called *cervical*, the next 12 *thoracic*, the next five or six are *lumbar*, five are fused to the *sacrum*, and what some refer to as your "tail bone"— made of three tiny vertebrae—is actually called the *coccyx*.

Ribs and Sternum

Twelve pairs of bones wrap around your chest to protect vital organs like the heart and lungs. The top seven connect directly to the sternum and are called *true ribs.* The others connect indirectly through the seventh rib and are called *false ribs.*

Pectoral Girdle

These bones connect the arms to the axial skeleton.

Pelvic Girdle

These bones connect the legs to the axial skeleton.

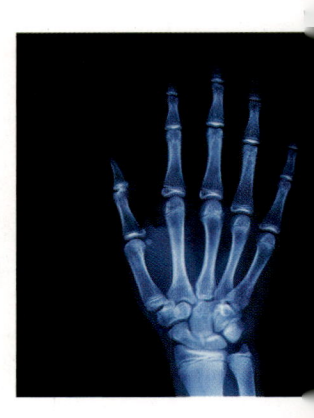

Wrist

Your wrist consists of eight small bones called *carpals* that allow precise and steady rotating of your hand.

Hand

Connected to the carpals of your wrist are the five metacarpals of your hand. Each metacarpal is connected to a finger or thumb. Your fingers consist of three bones called *phalanges*, while your thumb only has two.

Arm

The arm consists of the upper bone called the *humerus* and the bones of the forearm called the *radius* and *ulna*. These two bones exhibit masterful design in the way they allow your arm to rotate, giving you flexibility and stability in your wrists and hands.

Did you know?

When the Bible says God took a rib from Adam to form Eve, the Hebrew word translated as "rib" actually means "side." Even if "side" included Adam's rib, it would not imply men should have fewer ribs than women, since any loss of a body part would not be inherited by Adam's children.

Leg

The upper bone of your leg is called the femur, which is the longest bone in your body. The lower part of your leg is made up of the tibia and fibula. The tibia is larger than the fibula and bears most of your body's weight.

Ankle

The ankle and heel part of your foot consists of seven tarsals that allow your foot to move.

Foot

Connected to the tarsals are the five metatarsals of your foot. Each metatarsal is connected to a toe. Your toes consist of three phalanges, except for your big toe, which consists of only two. The arched shape of the foot acts as a spring to absorb the shock of walking.

Bone Categories

Long

These are bones with marrow on the inside, like the femur and humerus. Other long bones include the phalanges, tibia, fibula, radius, and ulna.

Short

These are often cubed or round. Examples include the tarsal bones of the foot and the carpal bones of the hand.

Flat

Flat bones vary greatly, but all are thin in some way. Examples include ribs, hip bones, and several bones in the upper part of the skull.

Irregular

These are bones so different from the rest that they defy description. The vertebrae and some bones of the skull are irregular bones.

Sesamoid

These are embedded in tendons and protect and reinforce the tendons at particular joints. The best known is the patella, or kneecap. Other sesamoids may grow at different joints, but only some people have them.

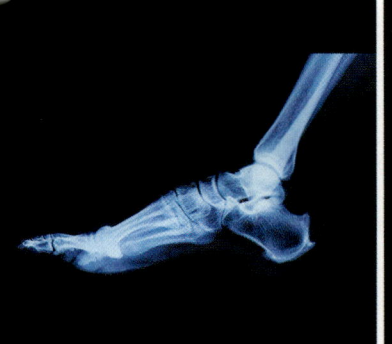
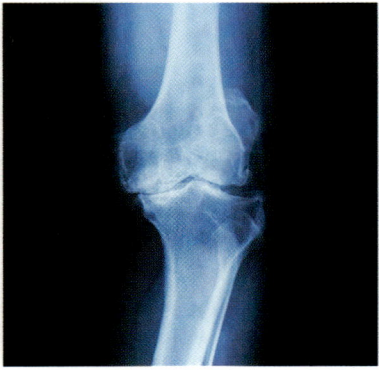
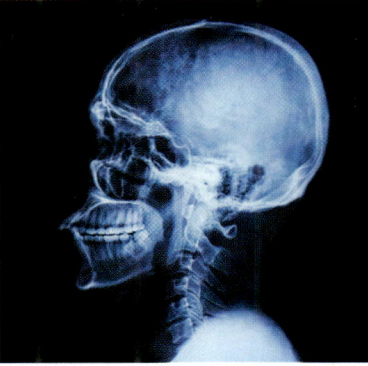
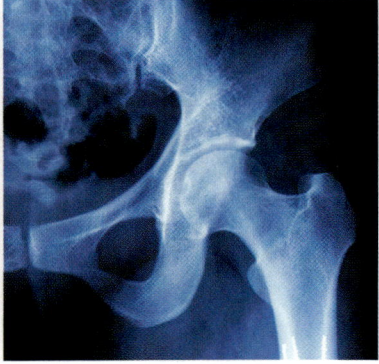

Bones: The Inside Story

If you were to design a framework for the human body, what materials would you use? Would you pick rubber to make it flexible? Or steel to make it sturdy? Well, God has gifted you with a framework of bones made of both harder and softer materials in just the right combination to achieve maximum efficiency, strength, flexibility, and endurance as you live and move. If your bones were hard all the way through, you'd have very limited motion like a stiff robot. But if bones were soft and flexible throughout, you'd be lying like a heap of spaghetti because your frame wouldn't hold up under your body's weight. The careful balance of strength and flexibility in the composition of bones is only one of the amazing characteristics of the human skeleton, showing God's wisdom in His creation.

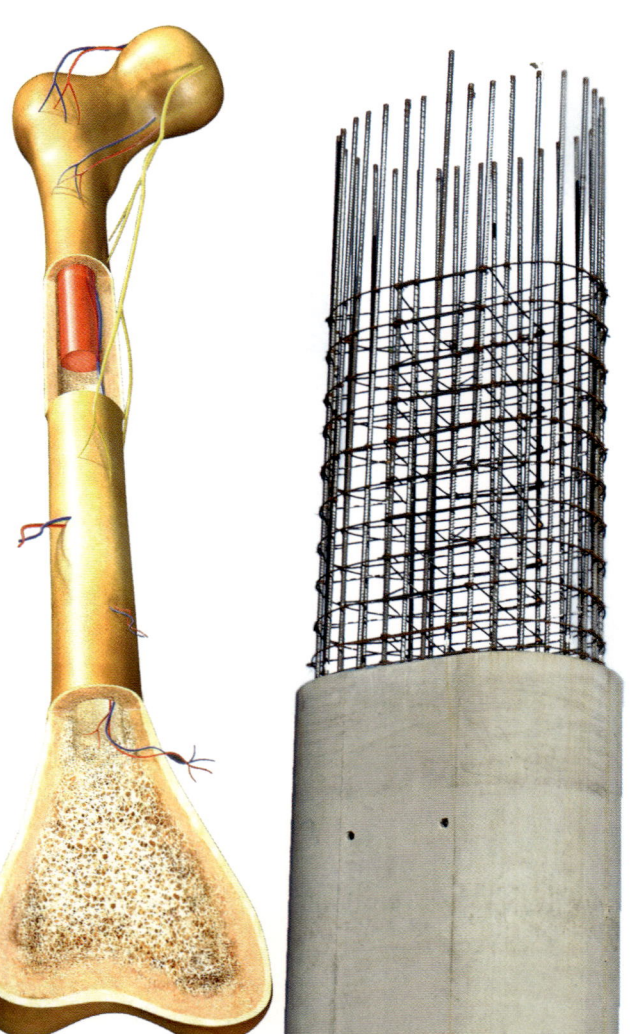

INSIDE BONE

The insides of bones are made of spongy bone that not only absorbs shocks but also braces the outer shell. It allows us to bend and twist more than if we just had the outer shells.

THE OUTER SHELL

The outer shell of bone is constructed a lot like concrete. The "cement" part is a substance called *apatite*, a medium-hard mineral with properties similar to marble. If you were to look at bone through a microscope, you would see that apatite is supported by a kind of "rebar" called *collagen*. These fibers are connected in a kind of mesh grid that is actually harder to pull apart than steel, giving bones flexibility and strength greater than concrete.

DID YOU KNOW?

360 collagen fibers could be put end to end in the width of a human hair!

REMODELING

Bone thicknesses change throughout a person's lifetime. This highly efficient process called *remodeling* ensures that more bone is built in specific locations where it needs to support heavier loads and less bone is built in places that don't bear much weight.

DID YOU KNOW?

100% of a baby's bones are replaced every year. For people in their 20s, the equivalent of 20% of the skeleton is replaced yearly—though high-stress areas like the top of the upper leg bone may be replaced up to three times per year.

FRACTURES

Bones also have ways of dealing with fractures. First, blood clots around the fracture and starts to repair it. Within 48 hours, other cells gather within the blood clot and use it as a template to build a microfiber meshwork that supports further repair work. The fracture zone is full of bone fragments and dead cells, things that are useless to the repair work, so cells engulf and digest the debris. Fibroblasts lay down collagen fibers over the blood clot that can span the break. Once some collagen bridges are laid down, new cartilage can be placed. Once built, this collagen-cartilage unit functions as a new kind of rebar for the fractured bone.

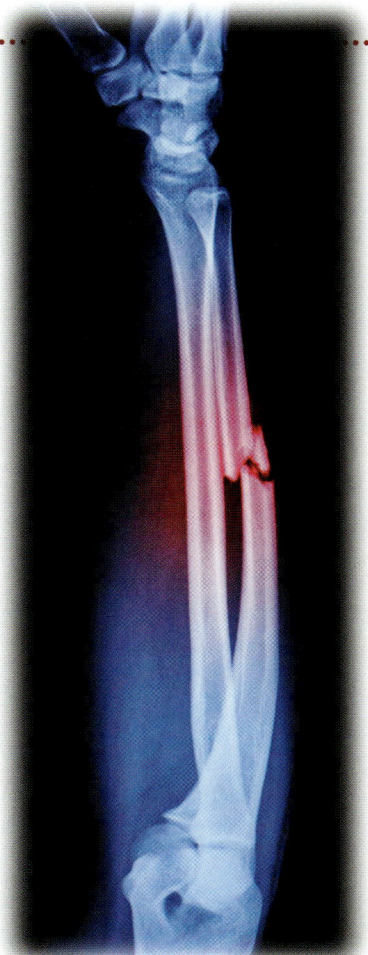

SACRIFICIAL BONDS

Automobile makers design "crumple zones" of material that fold up to absorb crash forces. Bones resist fractures in a similar way. Sacrifical bonds are microscopic weaker areas placed at regular areas within bones. They often break upon impact instead of the whole bone. Their exact arrangements in bone absorb and then disperse many forces that could otherwise reach the fracture threshold.

DID YOU KNOW?

If you break a bone in your arm or leg, your doctor will likely wrap it in a cast to keep the bones aligned while they heal. A cast has soft padding on the inside, with a stiff outer layer to hold the bone in place while the body repairs it. Casts were even used by ancient Egyptians, though they chose different materials to make them.

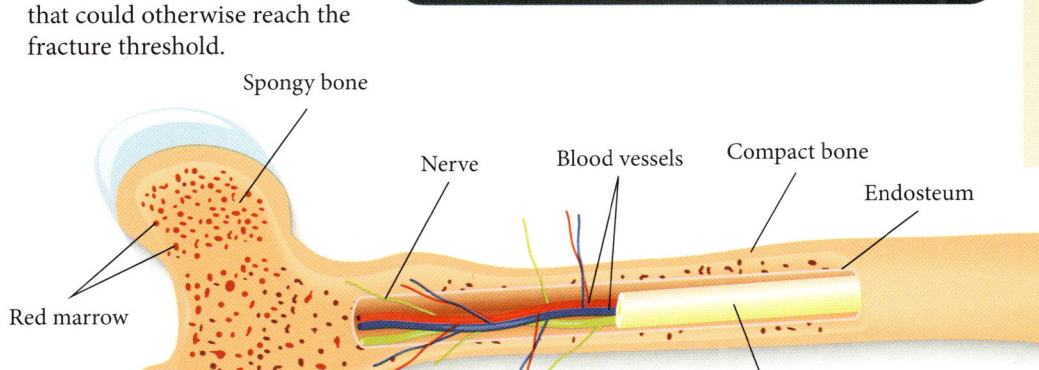

Spongy bone

Nerve Blood vessels Compact bone

Endosteum

Red marrow

Yellow marrow

Periosteum

BONE MARROW

Soft, fatty tissue that produces blood cells and platelets.

SPONGY BONE

Made of columns of bone tissue called *trabeculae* with red bone marrow between.

COMPACT BONE

Made of collagen fibers and hard mineral salts.

PERIOSTEUM

The outside layer of bone, composed of strong collagen fibers.

The Muscular System: Made to Move

Here's a riddle for you: What body parts use most of the energy you get from food and oxygen and make up almost half your body weight?

Hint: They get bigger by moving heavy objects, and the smallest ones are in your ear.

Give up? They're your muscles! Together they make up the muscular system, and their contractions help you live, move, eat, and breathe. Your body has hundreds of muscles, but all can be categorized in just three types: skeletal, cardiac, and visceral.

SKELETAL MUSCLES

Skeletal muscles are the ones people typically try to exercise. Your body has around 700, including the biceps and triceps in your arms, abdominal muscles in your torso, and quadriceps and calves in your legs. Skeletal muscles are the only kind that can be controlled with conscious thought, so everything you consciously do with your body requires skeletal muscles. All of them attach to the skeleton in at least one place. However, most of them attach to two different bones, so when they contract they pull the bones together and move your body. In any given motion you perform, the particular skeletal muscle that contracts is called the *agonist*, or "prime mover." An *antagonist* muscle operates on the opposite side of the joint to move it in the opposite direction. If you were to pick up a bowling ball, your biceps would flex as the agonist muscle. The antagonist muscle, the triceps, would contract when you straighten your arm.

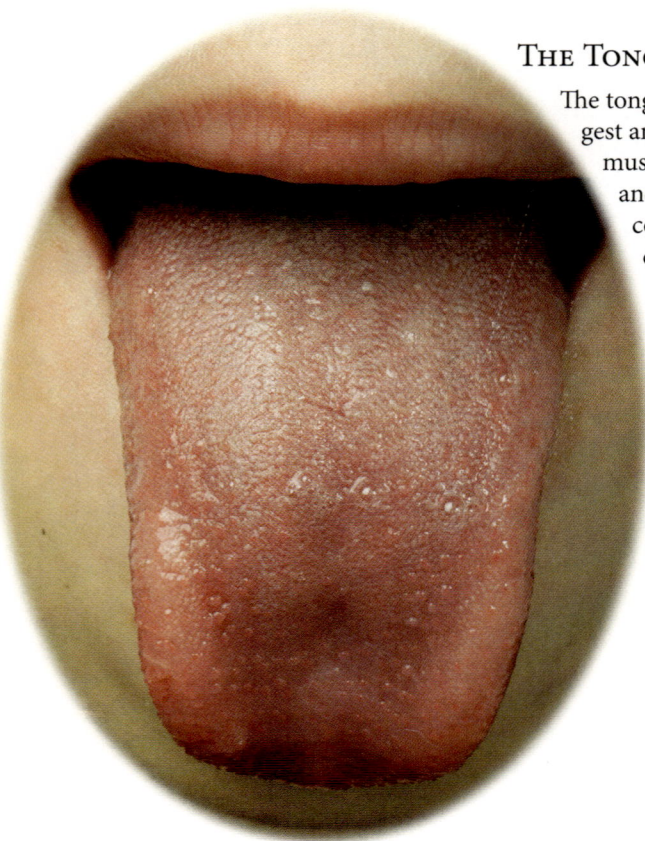

THE TONGUE

The tongue has some of the strongest and most flexible skeletal muscles. It helps you eat, digest, and speak. Your tongue is covered by a pink tissue called the *mucosa* that has hundreds of tiny bumps called *papillae*.

DID YOU KNOW?

The tongue is not a single muscle but is composed of eight muscles. Four originate in the neck, skull, and jaw. Four pairs of intrinsic muscles originate in the tongue itself and connect to connective tissue within the tongue.

CARDIAC MUSCLE

You only find cardiac muscle in one place: the heart. Like visceral muscles, the heart's movement is involuntary, but it can actually stimulate itself to contract. Nerve inputs and hormones can also influence heart rate. Your cardiac muscle is striated, having light and dark stripes when viewed under a microscope.

VISCERAL MUSCLES

Visceral muscles are completely involuntary and line organs such as the stomach, intestines, esophagus, uterus, respiratory passageways, and blood vessels. They help pass things through your digestive system, moving in wave-like movements. They are also called smooth muscles because they have a smooth appearance when viewed under a microscope, unlike the banded appearance of skeletal and cardiac muscles. Your body has so many visceral muscles that it's difficult to count them.

FATIGUE

After strenuous activity, muscles grow fatigued and lose their ability to contract. A fatigued muscle uses all of its nutrients, including oxygen, glucose, or ATP, and fills up with waste products like lactic acid and carbon dioxide. The build-up of lactic acid can make you feel very sore. A fatigued muscle has "oxygen debt," and your body must take in extra oxygen to make up for it, causing you to take in large, deep breaths.

MUSCLE ATROPHY

If you don't use your muscles, they can begin to waste away. This condition is known as *muscle atrophy*. Muscle atrophy, which is caused by a lack of physical activity, can often be reversed through proper diet, exercise, or physical therapy. Some diseases can cause muscle atrophy, such as amyotrophic lateral sclerosis (ALS, or Lou Gehrig's disease), which affects nerve cells that control voluntary muscle movement.

MOTOR NEURONS

Nerve cells known as *motor neurons* control the movements of skeletal muscles. Muscle movement occurs when your brain sends neurological signals to these motor neurons, telling them how many muscles to contract or expand according to the weight being lifted. Some muscles, such as leg muscles, have very few motor neurons since they do not require fine movements. Muscles that require more precision, such as those in your fingers and eyes, have many more motor neurons. The contractions that produce skeletal movement are called *isotonic contractions*. Contractions that do not produce movement are called *isometric contractions* and cause your body to become tense, like from stress or fear.

All together, our nerves, muscles, and skeleton work together in ways that allow the human body to perform some pretty amazing physical feats.

The Circulatory System: A Matter of the Heart

Your circulatory system is like an elaborate network of river canals used for transporting goods. It consists of the heart, lungs, and blood vessels. Blood picks up nutrients from the "port" of your liver, and oxygen from the "port" of your lungs, and delivers them throughout the body using the "canals"—known as blood vessels—of your circulatory system. Their rapid delivery service is aided by the pumping of your heart.

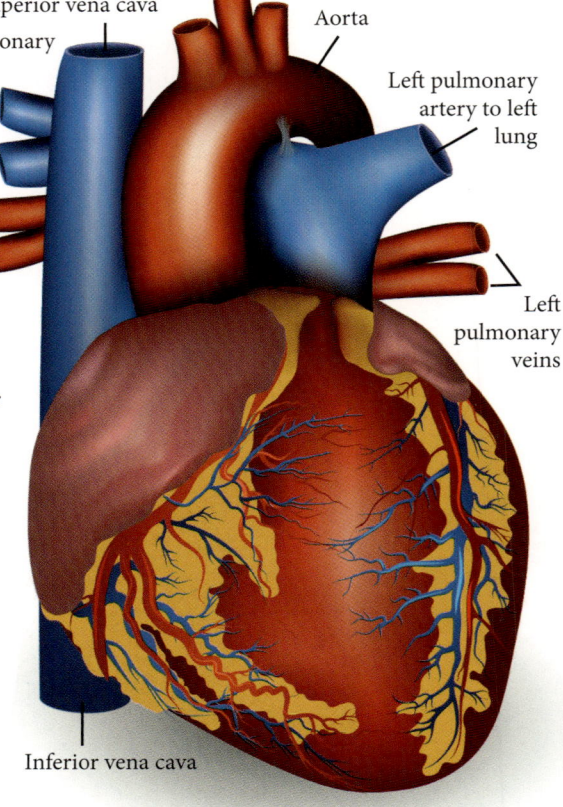

Superior vena cava

Aorta

Right pulmonary arteries to right lung

Left pulmonary artery to left lung

Right pulmonary veins

Left pulmonary veins

Inferior vena cava

THE HEART

Made almost entirely of pure muscle, the heart has four chambers. The right atrium and ventricle process oxygen-poor blood and send it to the lungs for replenishment. The now oxygen-rich blood is circulated to the left atrium and ventricle. The top of the heart is called the *base*. It connects the heart to the major vessels of your body: aorta, vena cava, pulmonary trunk, and pulmonary veins. A wall of muscle called the *septum* separates the right and left sides of the heart.

THE LUNGS

When the lungs inhale air, microscopic air sacs called *alveoli* absorb oxygen from the air and help transfer it into the blood. Blood vessels in the pulmonary circuit then carry the oxygen from the lungs to the heart.

DID YOU KNOW?

The average human body contains about five liters of blood, and your heart at rest can pump it all through your body every minute!

DID YOU KNOW?

If all your blood vessels were laid end to end, they would wrap around the earth more than twice!

BLOOD VESSELS

Blood vessels contain a hollow area called the *lumen* through which blood flows. Surrounding the lumen is the vessel's wall. There are three major types of blood vessels: arteries, capillaries, and veins.

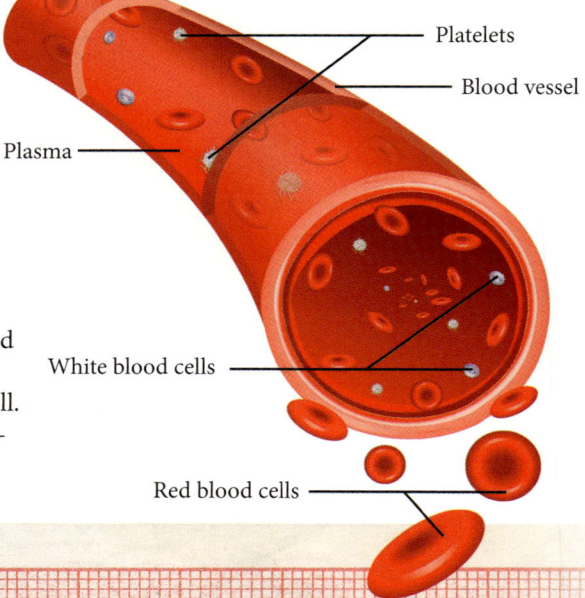

Platelets

Blood vessel

Plasma

White blood cells

Red blood cells

A medical machine called a *cardiograph* is often used by doctors to record the electrical signal moving through your heart. It displays them in the form of a cardiogram.

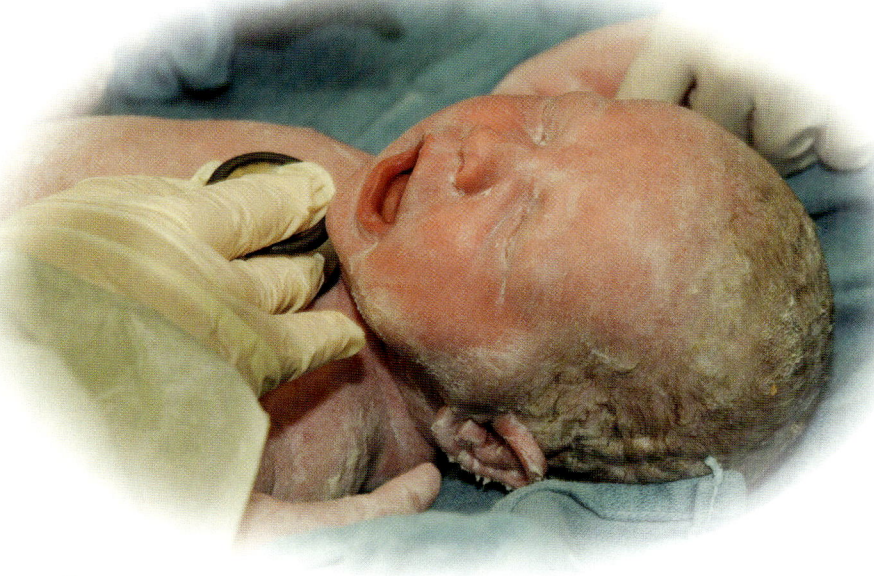

ARTERIES

Arteries are large-diameter blood vessels that carry highly oxygenated blood to all body tissues. The walls of the arteries are thicker, more elastic, and more muscular than other blood vessels since they handle more pressure from being closer to the heart. Some arteries are able to expand or contract according to how much blood flows from the heart. For example, when your body is trying to cool down and the vessels expand, they carry more blood to the limbs.

CAPILLARIES

Capillaries are the smallest and thinnest blood vessels and also the most numerous. They branch off from arterioles, which are smaller branches of arteries. Capillaries carry blood close to the cells of the body tissues to exchange gases, nutrients, and waste products. The walls of capillaries are very thin, making these exchanges as efficient as possible.

VEINS

Once blood has carried oxygen to the tissues, it returns to the heart through large veins. The heart then pumps this oxygen-poor blood to the lungs. *Venules* connect capillaries to veins. The veins and venules do not handle the same amounts of pressure as arteries, so they can be thinner, less elastic, and less muscular. They use gravity, inertia, and muscle contractions to push blood back to the heart.

TWO CIRCUITS

The *pulmonary circuit* consists of the blood vessels that leave the right side of the heart, pick up oxygen from the lungs, and then return to the left side of the heart to be pumped out to the rest of the body.

The *systemic circuit* consists of the blood vessels that leave the left side of the heart, bring oxygen to the whole body, and then return to the right side of the heart.

Life-Giving Blood

From the earliest days in the mother's womb until the day of death, a person's life is in the blood. Even a person-to-person gift of blood is treasured and called "the gift of life." Human blood is indeed a gift from the Lord Jesus Christ, clearly testifying to His great creative abilities and the body's total unity of function.

Blood is made up of oxygen-carrying red blood cells, disease-fighting white blood cells, and platelets suspended in a liquid that is 92% water, called *plasma*. Hematopoietic stem cells in your bone marrow are capable of transforming into red blood cells, white blood cells, and platelets. After approximately seven days of maturation, they are released into the bloodstream.

Blood's liquid form is perfectly suited for its purpose. Its quick-flowing capability enables it to transport vital substances throughout the body with ease. Blood and blood vessels form an incredibly flexible conduit, molding to the shape of your body at any given moment. Because of this, you can move and bend without hindering your blood's circulation.

"For the life of the flesh is in the blood." (Leviticus 17:11)

BLOOD TYPE

Everyone's blood is not the same. Red blood cells are marked with antigens. Your body's specific antigen markers determine your blood type (A, B, AB, or O). The presence (positive) or absence (negative) of the Rh antigen differentiates blood as well. Donated blood must be compatible with the blood type of the receiver or the body will destroy the donated cells.

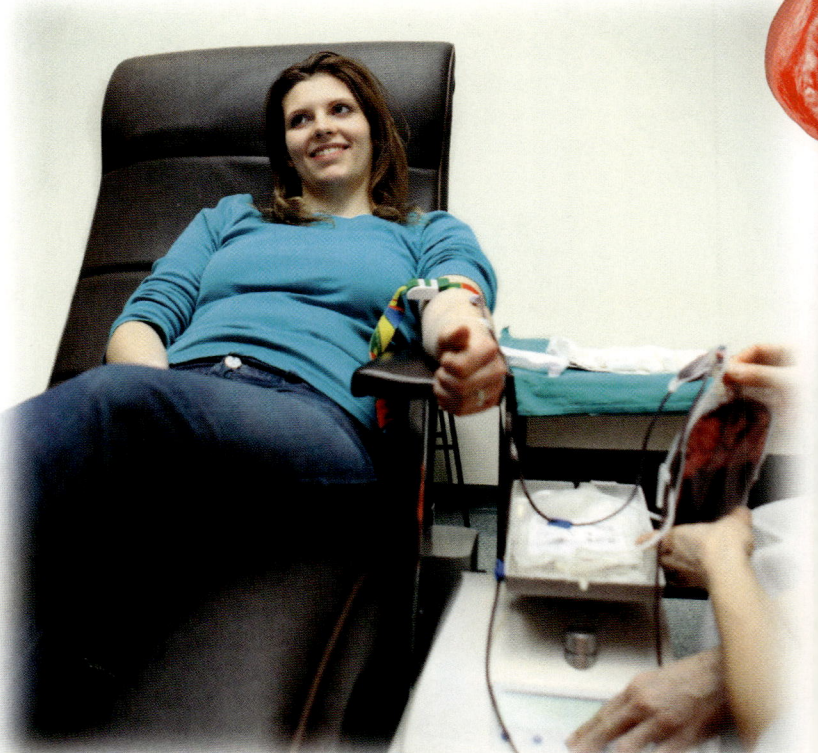

Red Blood Cells (Erythrocytes)

Red blood cells (RBCs) are the most abundant cell in blood. They contain the protein *hemoglobin*, which gives blood its red color. Hemoglobin helps carry oxygen from the lungs to the entire body and then returns carbon dioxide to the lungs so it can be exhaled.

A single red blood cell is shaped similar to a donut, but the middle is flattened instead of cut out. Both sides of the flattened center have a bowl-like indentation. This unique shape provides the highest possible membrane surface area for its size, enabling over 250 million hemoglobin molecules in each of the billions of RBCs to be oxygen-loaded or off-loaded in a fraction of a second.

Red blood cells are rather unique in that they don't have a nucleus. This makes them quite flexible, helping them fit through even the tiniest blood vessels in your body. However, the lack of a nucleus limits the life of the RBC as it travels through the body, damaging its membrane and depleting its energy supply. A red blood cell exists about 120 days.

White Blood Cells (Leukocytes)

White blood cells account for only about 1% of your blood but they serve a grand purpose as they regulate microbes, including infectious ones. The most common white blood cell is the *neutrophil*, the "immediate response" cell. Each neutrophil cell lives less than a day, so your bone marrow must constantly make new ones to maintain protection.

Another major white blood cell is a *lymphocyte*. Lymphocytes help regulate immune cells, attack infected cells, and make antibodies.

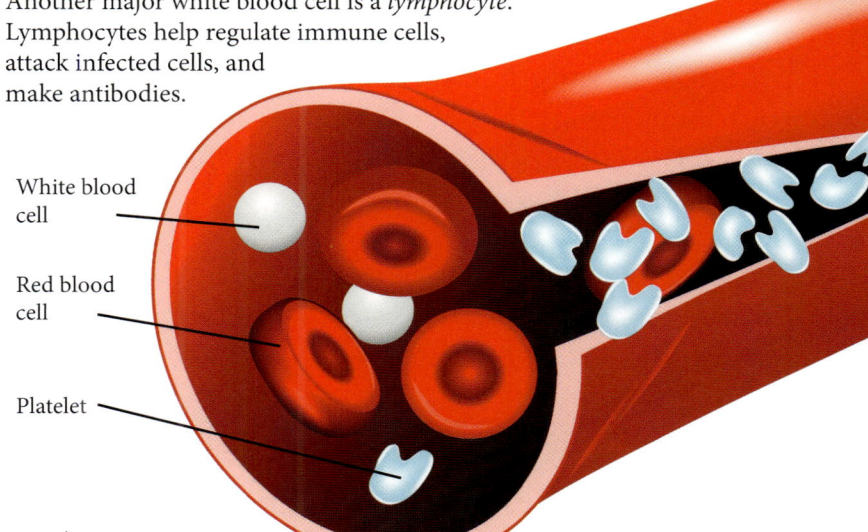

White blood cell

Red blood cell

Platelet

Platelets (Thrombocytes)

Unlike red and white blood cells, platelets are not actually cells but cell fragments. Platelets help the blood to clot by targeting the site of an injury, adhering to the lining of the injured blood vessel, and forming a platform on which blood clotting can occur. This results in the formation of a fibrin clot, which covers the wound and prevents further bleeding. When the clot hardens, it often forms a scab when it's on your skin.

An abnormally high number of platelets in the body can cause unnecessary clotting, leading to strokes and heart attacks. On the other hand, an extremely low platelet count can lead to extensive bleeding.

Plasma

Plasma transports blood cells throughout the body along with about 120 dissolved components necessary for daily function. Its cargo includes oxygen, glucose, albumin, waste products, antibodies, clotting proteins, hormones, and proteins that help maintain the body's fluid balance.

How Kidneys Cleanse Your Blood

The kidneys—two fist-size organs on either side of your spine—are part of the urinary system. They chemically extract excess water and waste products. The kidneys divert this liquid waste to the bladder to be temporarily stored as urine. When the bladder fills, it stretches. Stretch sensors then notify the brain of bladder pressure. Urine exits the body through a tube at the bottom of the bladder called the *urethra*.

The Bible says that the Lord Jesus' blood is particularly special—in fact, "precious" (1 Peter 1:19)—because it is able to redeem us and cleanse us from all sin (1 John 1:9). His blood is truly life-giving to those who accept His gift of salvation.

The Respiratory System: A Breath of Fresh Air

"And the LORD God formed man of the dust of the ground, and breathed into his nostrils the breath of life; and man became a living being." (Genesis 2:7)

Relax and take a deep breath. Feel the air enter through your nose or mouth. Notice your chest expanding as your lungs inflate. Now exhale slowly. Multiple muscles and organs are adjusting as warm, humid air exits the same way fresh air came in.

You may be surprised to find that you'll repeat this process another 23,000 or so times today—usually without even thinking about it! The airway, lungs, and respiration muscles work together within the respiratory system to remove carbon dioxide and bring oxygen to every cell. Oxygen serves as a vital component of energy production for the body. Carbon dioxide is the waste product of this process, and your lungs release it as you breathe out.

NOSE AND NASAL CAVITY

Made of cartilage, bone, muscle, and skin, the nose serves as the main external opening of the respiratory system. The nasal cavity just behind the nose is lined with hairs and a mucus membrane that warm, moisturize, and filter air as it enters the body. They help trap dust, mold, pollen, and other environmental contaminants before they can get to the inner parts of the body. Most normal breathing occurs through the nose.

MOUTH

You can breathe through your mouth, but the pathway to the lungs is shorter so it doesn't warm and moisturize the air as much as breathing through the nose. The mouth also doesn't remove as many contaminants since it doesn't have hairs and mucus like the nose. However, inhaling through your mouth helps if you're short of breath and need to fill your lungs quickly.

PHARYNX

This space connects the nasal cavity to the place where the esophagus and trachea converge.

EPIGLOTTIS

This flap of elastic cartilage is functionally part of the larynx and works like a mobile cover over the top of the entire trachea. Reflex muscles work together to drop the epiglottis toward a horizontal position and pull the larynx upward against it tightly when you swallow. It covers the trachea and allows food to be pushed down your esophagus.

DID YOU KNOW?

Plants need the carbon dioxide you exhale in order to grow. Returning the favor, they release oxygen, the very thing we need to survive—a perfectly designed relationship!

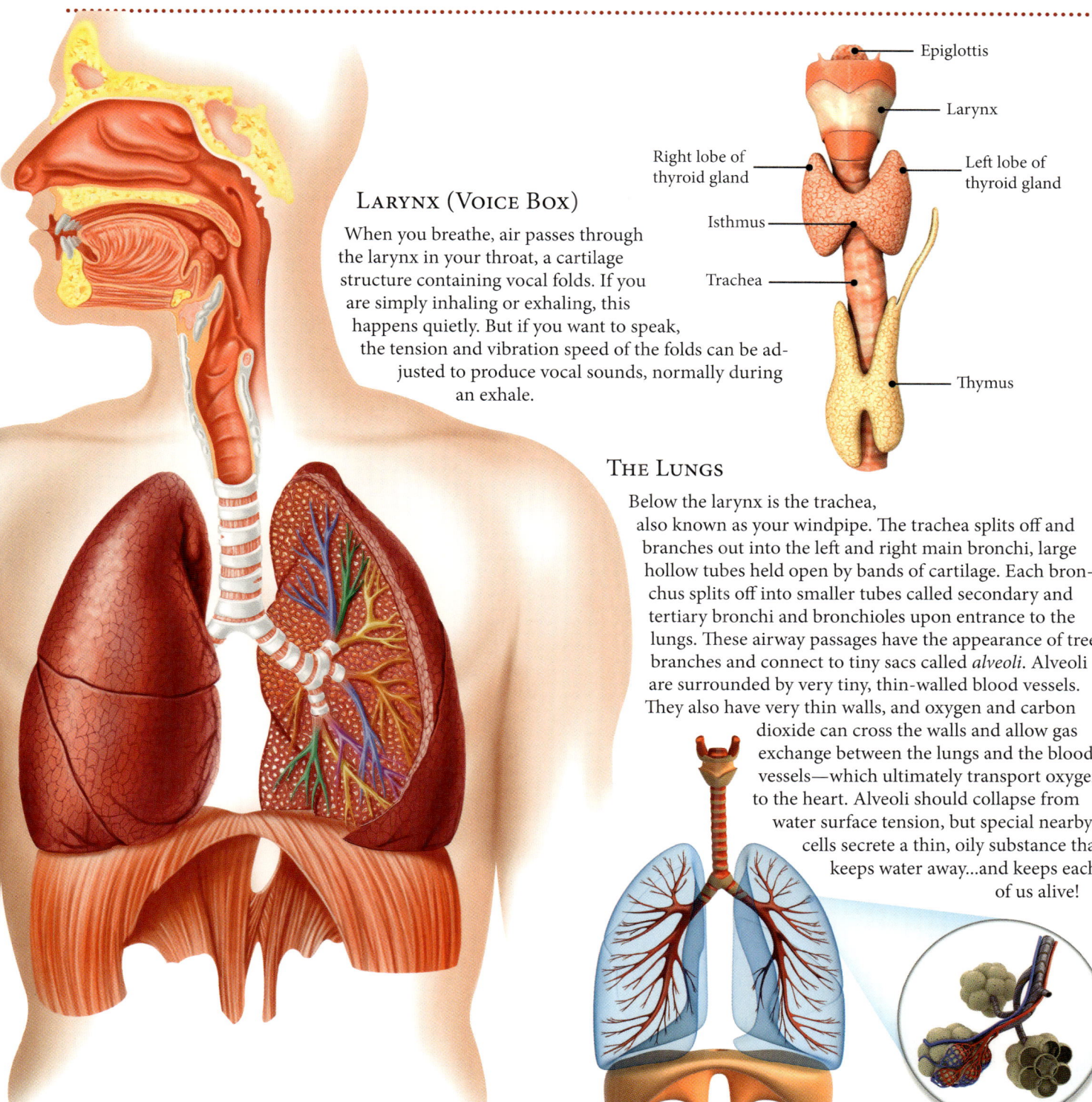

Epiglottis

Larynx

Right lobe of thyroid gland

Left lobe of thyroid gland

Isthmus

Trachea

Thymus

Alveoli

Larynx (Voice Box)

When you breathe, air passes through the larynx in your throat, a cartilage structure containing vocal folds. If you are simply inhaling or exhaling, this happens quietly. But if you want to speak, the tension and vibration speed of the folds can be adjusted to produce vocal sounds, normally during an exhale.

The Lungs

Below the larynx is the trachea, also known as your windpipe. The trachea splits off and branches out into the left and right main bronchi, large hollow tubes held open by bands of cartilage. Each bronchus splits off into smaller tubes called secondary and tertiary bronchi and bronchioles upon entrance to the lungs. These airway passages have the appearance of tree branches and connect to tiny sacs called *alveoli*. Alveoli are surrounded by very tiny, thin-walled blood vessels. They also have very thin walls, and oxygen and carbon dioxide can cross the walls and allow gas exchange between the lungs and the blood vessels—which ultimately transport oxygen to the heart. Alveoli should collapse from water surface tension, but special nearby cells secrete a thin, oily substance that keeps water away...and keeps each of us alive!

Abdominal Wall

The main muscles used for exhaling make up the abdominal wall, commonly called the "six pack." They control the exhalation of air when you sing, speak, or cough.

Did you know?

The left lung is slightly smaller than the right to accommodate the position of the heart. The two lungs together can hold around four to six liters of air.

Diaphragm and Intercostal Muscles

The main muscle of respiration is the diaphragm, a thin sheet of skeletal muscle underneath the lungs that attaches to the lower ribs. Small intercostal muscles stretch between the ribs to form the chest wall. When you inhale, the diaphragm moves down and intercostal muscles pull up and out to expand the chest cavity and inflate the lungs.

The Digestive System: Harnessing Energy for Life

If we ever doubt our need for food, it only takes one or two missed meals to remind us of its significance. But we probably don't often consider the processes that take our food and convert it into the energy we need for life. How does eating things like pizza, carrots, cereal, or apples empower us to move our bodies?

When it comes to the digestive system, our body gets to work before we even take a bite. As the brain senses the smells, sights, or even the sounds of food, it starts to prepare the mouth to eat by secreting saliva. That's why your mouth "waters" around food. When food enters the mouth, teeth chew it up into small pieces and saliva helps to break down the chemical makeup of the food and pack it into a sticky wet ball called a *bolus* that can be easily swallowed.

When swallowed, the bolus enters the esophagus while the brain sends signals to two different muscles. The muscle above the bolus contracts and the muscle below relaxes. This wave-like movement squeezes the bolus down the esophagus to your stomach with a force strong enough to do this even if you're upside down!

The stomach acts as a mixer and grinder. Composed of elastic fibers and three layers of muscles, it churns the bolus while over a dozen enzymes, hormones, and other chemicals break down the food into a fine paste.

DID YOU KNOW?

If it did not have a mucous lining to protect it from the hydrochloric acid used to break down food, your stomach would digest itself!

WHY DO WE FEEL HUNGRY?

Our digestive systems are designed to anticipate when we will eat. At our usual meal-times, the stomach fills with acid. If no food enters the stomach, the acid causes hunger pangs. It takes about a week to reprogram the system's timing by eating on a different schedule.

Many of the nutrients your body receives from digesting food will be used to repair and build body tissues like bone, but about 60% will be used simply to keep your body running. At rest, the average person will need to generate the same energy needed to power a 120-watt light bulb. Heavy work can drive up the power demand tenfold.

One example of the chemical energy your body gains from food is the glucose sugar molecule. Your body uses the electrons stored in glucose to make a substance called *adenosine triphosphate* (ATP), which is fuel for the body, similar to burning coal for electrical energy.

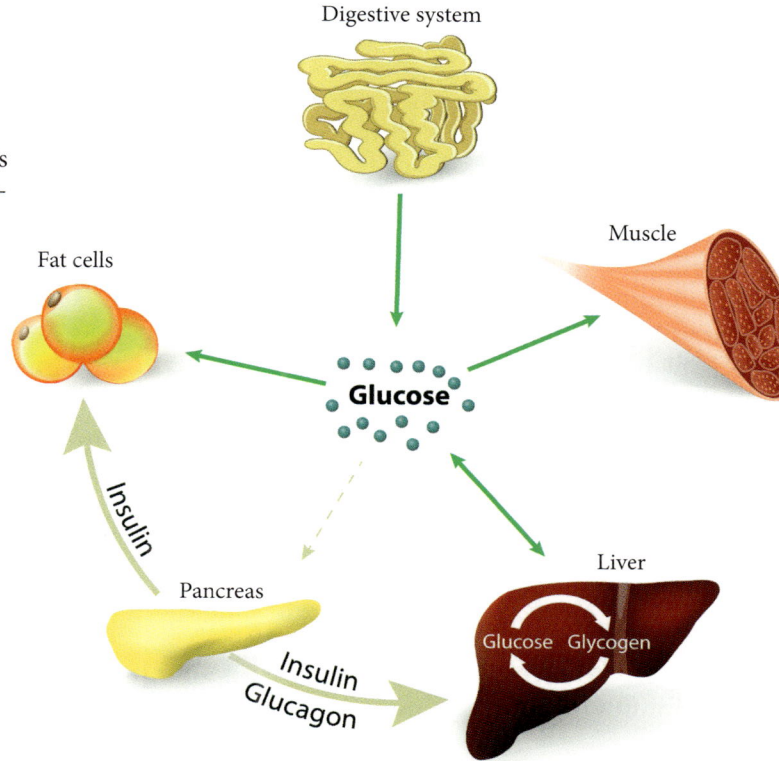

Digestive system

Fat cells

Muscle

Glucose

Insulin

Pancreas

Insulin
Glucagon

Liver

Glucose Glycogen

Glucose is the main energy source our digestive systems extract from food.

The pancreas produces enzymes that the duodenum uses to digest food.

The liver makes bile for the digestive system and is also essential to six of the body's systems.

The gall bladder stores bile made by the liver. It injects the bile into the small intestine to make fats and oils digestible.

Broken-down food enters the small intestine, where nutrients are absorbed and transferred to the blood.

Next, the waste enters the large intestine, also called the colon. By this time, the body has grabbed all the nutrients it can from the food, leaving remnants of food debris and bacteria known as *stool*. After passing through the colon to remove water, the stool will finally reach the rectum about 12 to 36 hours after the food was eaten.

The anus consists of muscles arranged in a ring called a *sphincter*. This open-close valve keeps the stool in the rectum until the brain decides that it can be emptied.

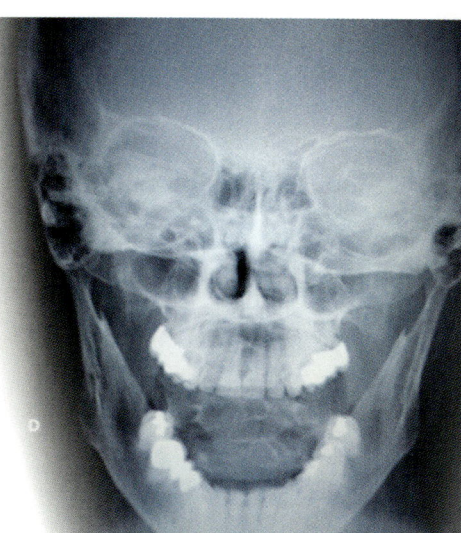

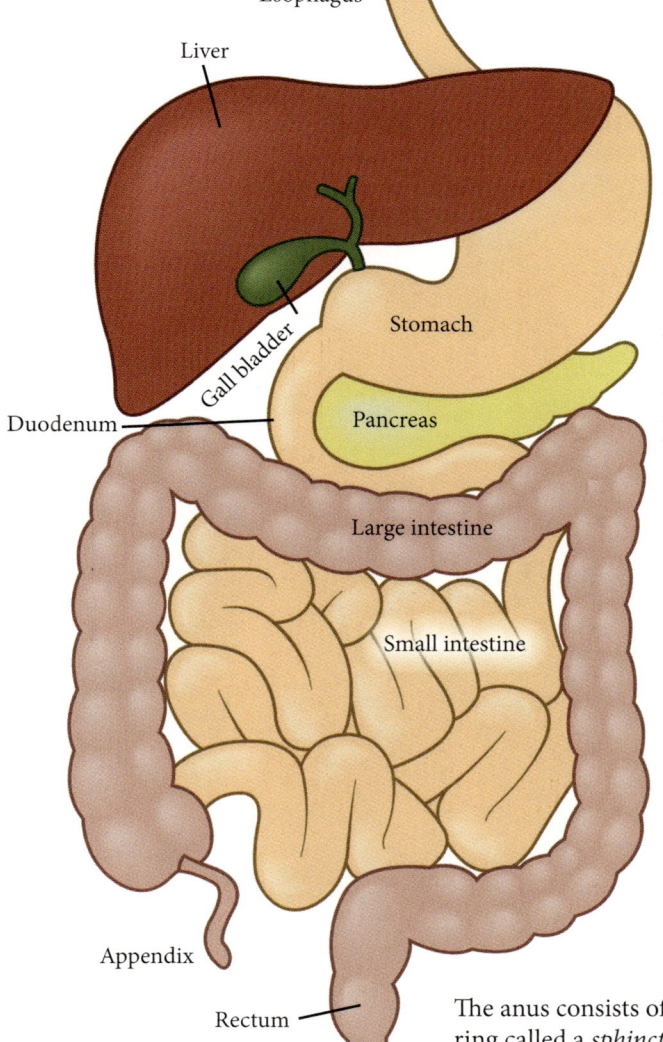

Esophagus

Liver

Gall bladder

Stomach

Pancreas

Duodenum

Large intestine

Small intestine

Appendix

Rectum

Anus

DID YOU KNOW?

A normal meal takes about four hours to break down, while a high-fat meal can take six hours or more.

Food and What Our Bodies Do with It

Food could be just a boring and tasteless necessity, but God has shown His creativity and goodness by creating a variety of colors, flavors, and textures for the body to enjoy. A varied and healthy diet nourishes every cell, gives us energy, and prevents disease.

FOOD GROUPS

The U.S. Department of Agriculture used to make dietary recommendations using a diagram called the Food Pyramid. The pyramid has since been replaced by the image of a dinner plate, but it still focuses on proper proportions of the five food groups during meal times. The five main food groups are vegetables, fruits, grains, dairy, and protein. The carbohydrates, proteins, fats, oils, vitamins, and minerals found within these groups are essential for good nutrition.

DID YOU KNOW?

Jesus' 40-day fast in the wilderness (Matthew 4:1-2) was near the upper limit of how long a person can go without food.

DID YOU KNOW?

Water is just as essential for us as food because it is in every cell, tissue, and organ in our bodies. Water acts as a joint lubricant, regulates our body's temperature, and helps to digest food and flush away waste. You can only live about one week without water!

FROM SUNLIGHT TO BODY POWER

When people eat food, they are essentially ingesting "energy units." Marvellously designed photosynthetic machinery in plants captures sunlight, converts it into chemical energy, and stores that energy in edible forms such as leaves, fruits, and vegetables. When digested, equally marvelous machines in our bodies metabolize the food into energy for the body. When we eat meat, we get the energy units the animal gained from plants and stored in its tissues.

CARBOHYDRATES

Carbohydrates are the most important source of energy for your body. Their building blocks are sugars. You'll find carbohydrates in many foods: fruits, vegetables, breads, cereals, grains, and milk products. Foods that contain added sugars like cakes and cookies contain carbohydrates too, but it is important to limit these treats since too many of them can lead to obesity and health problems.

PROTEIN

Proteins play a role in every cell in your body. They are constantly being used, so it's important to replace them with the food you eat. The building blocks of proteins are *amino acids*. Humans need eight essential amino acids. Sources of protein include meat, poultry, fish, beans and peas, tofu, eggs, nuts and seeds, milk and milk products, grains, some vegetables, and even some fruits.

FAT

The broad category of fats is called *lipids*. Long chains of carbon and hydrogen atoms make up lipids. They can be more consolidated like the fat in meat or liquid like in oils. Healthy dietary lipids provide your body with the essential fatty acids needed to protect your heart, joints, pancreas, mood stability, and skin. Including good oils in your diet is vital because they support many of your body's functions and enable certain vitamins to be absorbed. Healthy sources include avocados, eggs, olive oil, nuts, and fatty fish. One example of an unhealthy fat is the man-made trans-fat often found in fried foods.

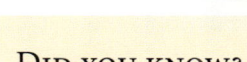

DID YOU KNOW?

The dark rim just below an avocado's peel contains antioxidants called *carotenoids* that help reduce tissue damage. The healthy fat deeper within the fruit increases your body's ability to absorb the carotenoids, making the avocado an ideal contribution to a healthy diet.

NUTRIENT ABSORPTION

The foods we eat must be broken all the way down to their building blocks. This process starts in the mouth. After the food passes through the stomach, it moves to a part of the small intestine called the *duodenum*, which adjusts the rate of digestion and combination of enzymes released based on the type of food you have eaten. This customized process optimizes nutrient absorption. The final products are single molecules of glucose, amino acids, and glycerol broken down from carbohydrates, proteins, and lipids respectively. These are distributed by the blood and lymphatic systems to all the cells in the body.

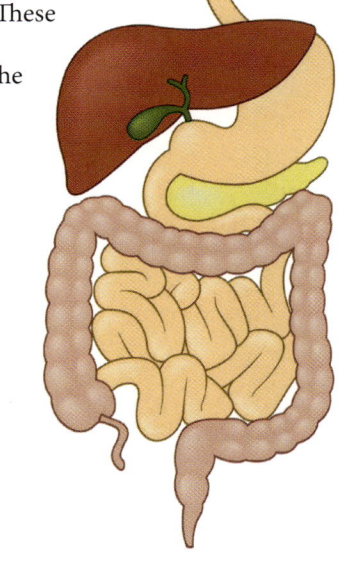

VITAMINS AND MINERALS

Vitamins and minerals are vital for proper growth and nutrition. You can usually get all you need by eating a variety of foods. The 13 vitamins your body needs are A, B1-B3, B5-B7, B9, B12, C, D, E, and K.

Macrominerals, minerals your body needs in larger amounts, are calcium, phosphorus, magnesium, sodium, potassium, and chloride. Trace minerals—iron, copper, iodine, zinc, fluoride, and selenium—are only needed in small amounts.

DID YOU KNOW?

If all of the folds and microscopic villi were smoothed out flat, the area of an average person's small intestine would cover over 1,500 square feet. This enables the body to capture nearly 100% of all potentially absorbable nutrients.

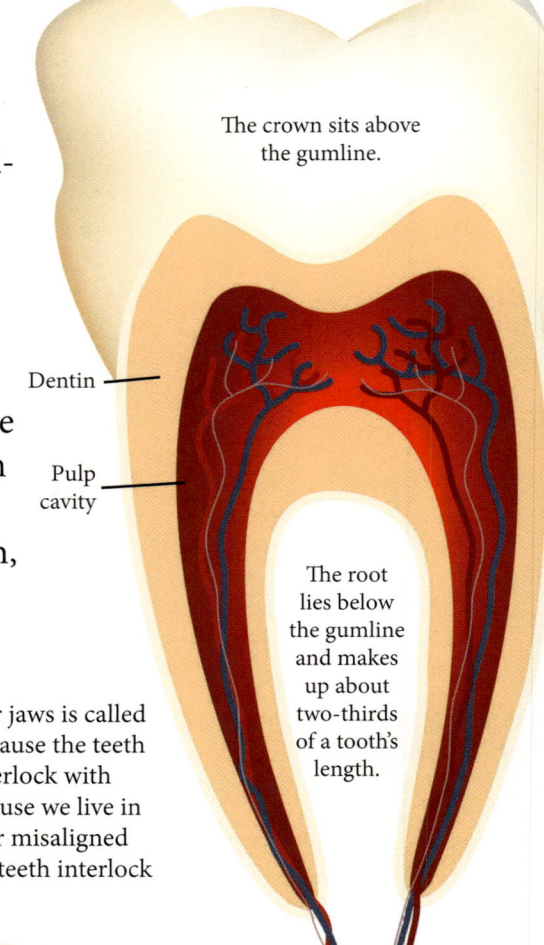

Enamel

The crown sits above the gumline.

Dentin

Pulp cavity

The root lies below the gumline and makes up about two-thirds of a tooth's length.

The Truth about Teeth

Have you ever imagined what life would be like without your teeth? People who have lost their teeth understand the difficulty of biting into a crisp apple or tearing meat off a drumstick without these important bony structures. God designed teeth to help break food down into pieces small enough to swallow and digest. Baby teeth are formed during development in the womb, but they usually do not begin to erupt (grow in) until somewhere between four and seven months after birth. We see God's amazing design for teeth when they automatically begin to fall out when we are around six or seven years old to make room for full-grown adult ones. The average adult has 32 teeth, most of which have grown in by about age 13.

OCCLUSION

The way our teeth fit together when we close our jaws is called *occlusion*. How could mutations just happen to cause the teeth in the upper jaw to have the right grooves to interlock with the teeth in the lower jaw—for all 32 teeth? Because we live in a fallen world, sometimes teeth are misshaped or misaligned and need orthodontic intervention, but the way teeth interlock shows God's design.

INCISORS (8)

These are the front teeth meant for shearing off pieces of food. They are the sharpest teeth in your mouth.

CANINES (4)

These are just outside the incisors. They have long roots and are meant for tearing food.

PREMOLARS (8)

These are located just behind the canines. They have a flat surface for chewing and crushing food.

MOLARS (8)

These are the back teeth, the flattest and biggest teeth to chew and grind food better than the premolars.

WISDOM TEETH (4)

These are in the very back and help grind food like the molars do. Wisdom teeth grow in later than the rest—at around age 18. However, some subpopulations of people have less robust jaws in response to a softer diet. So their wisdom teeth tend to displace the other teeth and are often surgically removed.

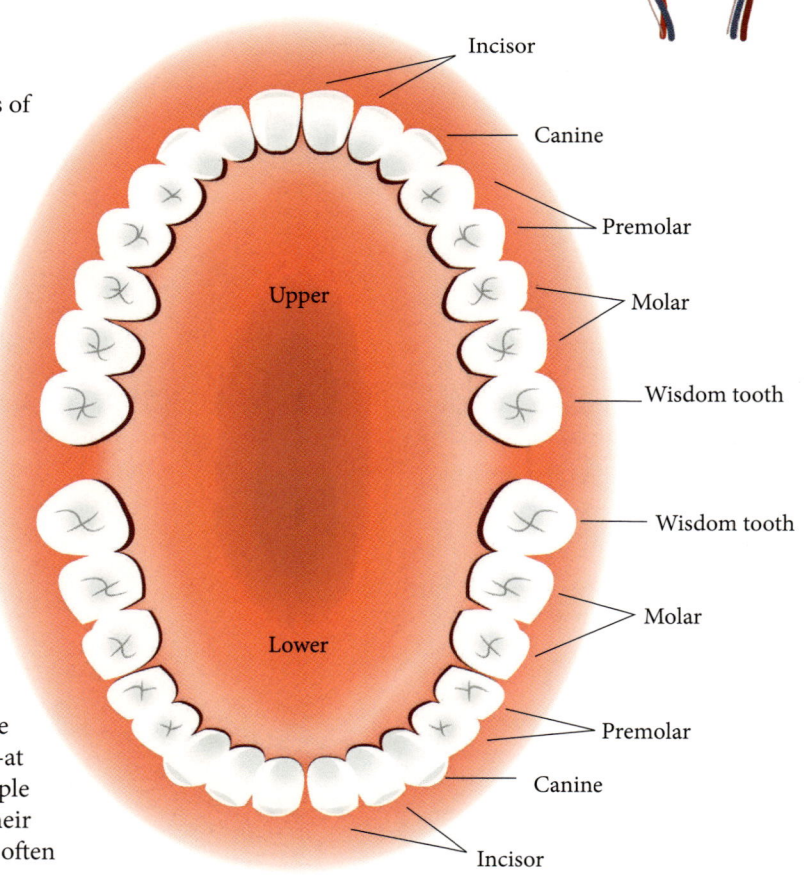

Incisor

Canine

Premolar

Molar

Upper

Wisdom tooth

Wisdom tooth

Molar

Lower

Premolar

Canine

Incisor

ARE WISDOM TEETH USELESS?

Were you ever told that your wisdom teeth had to come out because they were crowding your other teeth—and this was because the modern human jaw was slowly evolving? Actually, processed and refined foods in present societies have contributed to reduced jaw size. When refined foods are constantly eaten, there is less resistance to stimulate full jaw development, so erupting wisdom teeth crowd the other teeth. Failure of the wisdom teeth to erupt in normal positions can cause problems. The fossil record documents excellent, well-designed tooth patterns found in whatever creations are unearthed—be they people or dinosaurs. If evolution were true, museum display cases should be full of skulls clearly showing evolution's botched dental experiments—until the assumed processes of natural selection and mutations finally "got it right." But each time wisdom teeth—or any other type of teeth—are found in the fossil record, they're beautifully formed and ready for use.

TEETH TISSUES

Enamel covers the outside of the crown and is the hardest part of the tooth. Dentin is a bone-like material that supports, and is softer than, the enamel. It has nerve fibers that send pain signals when something's wrong with your tooth. The pulp is a soft tissue in the center of the tooth that contains blood, lymph vessels, and nerves. Cementum covers the root of the tooth and it, along with the periodontal ligament, attaches the tooth to the jawbone.

TAKING CARE OF YOUR TEETH

Simply brushing and flossing your teeth can protect them against all kinds of erosion and infections. Eliminating plaque—a sticky film that promotes bacteria—before it hardens into tartar is important for keeping your teeth healthy and pain-free.

CAVITIES

When plaque builds up, it can become a harder substance called *tartar*. Plaque can be brushed off, but tartar must be professionally removed. If you don't take care of your teeth, a build-up of plaque and tartar can dissolve the enamel and expose the sensitive dentin beneath. This decay of your teeth causes cavities. If a cavity gets too deep, it will cause pain and will have to be professionally treated.

DIET

Having a healthy diet is just as important as brushing and flossing, so keep sugary foods to a minimum. Sugar spreads around your teeth and provides an acidic environment for plaque-loving bacteria to thrive in.

The Nervous System: A Communication Network

The primary control center of the body is located in the central nervous system. The nervous system is like an intricate web of telephone lines that enables your brain to "talk" with different parts of your body from its home in your head. These communication networks are made up of nerves and specialized cells called *neurons*. All work together to allow information and brain commands to be transmitted throughout the body at an incredibly rapid rate.

NERVES AND NEURONS

Though nerves and neurons sound similar and are closely connected in their purpose, they are actually two different components of the nervous system. Neurons are specialized cells that communicate with other cells through connecting fibers called *axons*. A cylinder-shaped bundle of those axons make up your nerves, starting at the brain and central spinal cord and branching out to every organ, bone, and muscle in your body.

Glial cells, from the Greek word for "glue," surround nerve cells and feed, support, and protect them.

Cell body

Dendrites

Axon

COMMUNICATION WITHIN THE NERVOUS SYSTEM

As information and brain commands travel from neuron to neuron, they run into tiny gaps between axons called *synapses*. To keep the message from getting dropped at these gaps, chemical neurotransmitters are released to help pass the command on to the next neuron and ultimately to the destined region of the body. The journey of a command from the brain, through many nerves and neurons, and across synapses to arrive at the destined body part typically takes only a fraction of a second.

DID YOU KNOW?

Schwann cells insulate axons from random electrochemical interference. Young children have uncoordinated movements because their Schwann cells are not yet in place.

DID YOU KNOW?

A spinal tap is a procedure that places a needle into the space of the spinal canal below the level of your spinal cord to drain a sample of cerebral spinal fluid that is tested for infection or diseases.

CENTRAL AND PERIPHERAL NERVOUS SYSTEMS

The nervous system is divided into two main categories: the *central nervous system* and the *peripheral nervous system*. The two systems work together to enable us to think, move, and function.

The brain and spinal cord form the central nervous system—a control center that is constantly evaluating incoming information and deciding how to respond.

Everything else—the network of sensory nerves, clusters of neurons called *ganglia*, and sense organs—makes up the peripheral nervous system. This system constantly monitors every part of the body and sends information back to the brain. The peripheral nervous system connects the central nervous system to all the organs, as well as to the arms, hands, legs, and feet.

DID YOU KNOW?

Spinal cord injuries can cause loss of feeling and function to all parts below the level of injury. Paralysis from the waist down is called *paraplegia*. Paralysis from the shoulders down is referred to as *quadriplegia*.

SOMATIC AND AUTONOMIC SYSTEMS

Functionally, the nervous system has two main subdivisions: the *somatic* (voluntary) nervous system and the *autonomic* (involuntary) nervous system.

The somatic system consists of two kinds of neurons that connect the brain and spinal cord with muscles and sensory receptors in the skin. Afferent (sensory) neurons send sensory information to the brain, like the burning sensation you feel when touching a flame. The efferent (motor) neurons convey the necessary command from the brain to the muscles, telling your hand to move away from the flame. Motor neurons can also be autonomic.

The autonomic system connects the central nervous system to the organs and regulates specific body processes automatically, such as heart rate, blood pressure, digestion, and breathing rate. These vital processes continually function without your having to think about them.

NERVE DAMAGE

Different types of nerves can be damaged by disease or a spinal cord injury. If motor nerve fibers are destroyed, you may not be able to lift your arm because the command can't be transmitted from the brain to the arm muscles. If sensory fibers are injured, you may lose feeling in your hand because the information cannot get to the brain. Often combinations of both types of fibers are damaged in a spinal cord injury.

DID YOU KNOW?

There are around 100 million people in the United States who live with chronic pain. If you feel pain, it means that your central nervous system thinks your body is under threat and is trying to warn you about it. Pain is your brain's response to this signal.

A Beautiful Mind

Introducing the ultimate design challenge! Your job is to create a camera, a vast library, a complex computer, and an elaborate communication center, all in one. Your invention must constantly analyze new data, categorize and store images, supervise numerous tasks, transmit countless messages, and perform vital functions in life-or-death situations. It can never run out of memory. And by the way, it can't weigh much more than three pounds. Ready, set, go!

Sound impossible? Actually, it's already been done. The human brain is more advanced than any technology mankind can produce—nothing man-made can match the human brain in capacity, speed, efficiency, and learning ability. Our brains simply display the greatest concentration of order and complexity in the universe.

NEURONS

The fundamental component of the brain is the neuron, or nerve cell. Your brain contains about 100 billion neurons—sometimes called *gray matter*—and these are linked together by trillions of special connecting junctions called *synapses*.

Each tiny neuron contains a nucleus and branching slender extensions called *dendrites*. When a cell "fires," it sends an electrochemical impulse to its neighboring neurons through the dendrites. These firings make up our brain-wave patterns. Each neuron networks with around 10,000 other neurons, and the total number of connections in the human brain are almost beyond counting.

DID YOU KNOW?

Every single neuron in your brain contains a trillion atoms. Each neuron is like a microscopic universe, complete with order, purpose, and interdependent components.

DID YOU KNOW?

During the first nine months of life, neurons form at the astounding rate of 25,000 per minute!

MASTER OF DESIGN

The Lord Jesus Christ has mastered the ultimate design challenge. The human brain has remarkable purpose and interdependence within itself—every part working for the benefit of the whole body. Its awe-inspiring complexity contrasts with the simplistic secular attempts to explain the brain as a product of meaningless chance processes.

Knowing God designed the human brain should prompt us to use our minds to understand His purpose for His creation. Johann Kepler, a creationist and one of the greatest scientists, described science as "thinking God's thoughts after Him." We can look for God's thoughts, too.

Cerebrum

The largest part of the human brain is the *cerebrum*, Latin for "brain." The *cerebral cortex,* the outside and largest part of the cerebrum, is divided into four lobes. More complex mental tasks are performed by these lobes. Visual processing occurs at the back of the skull in the *occipital lobe.* The *temporal lobe* processes auditory information, sensory input, and memory. The *parietal lobe* integrates input from our various senses and manages spatial orientation and navigation.

The *frontal lobe* is the part of the brain that controls emotions, problem solving, language and speech, logic and judgment, and social skills. This lobe strongly influences our personality.

The brain stem connects the cerebrum to the spinal cord. Underneath the cerebrum is the *cerebellum*, which regulates muscle activity, coordination, and balance.

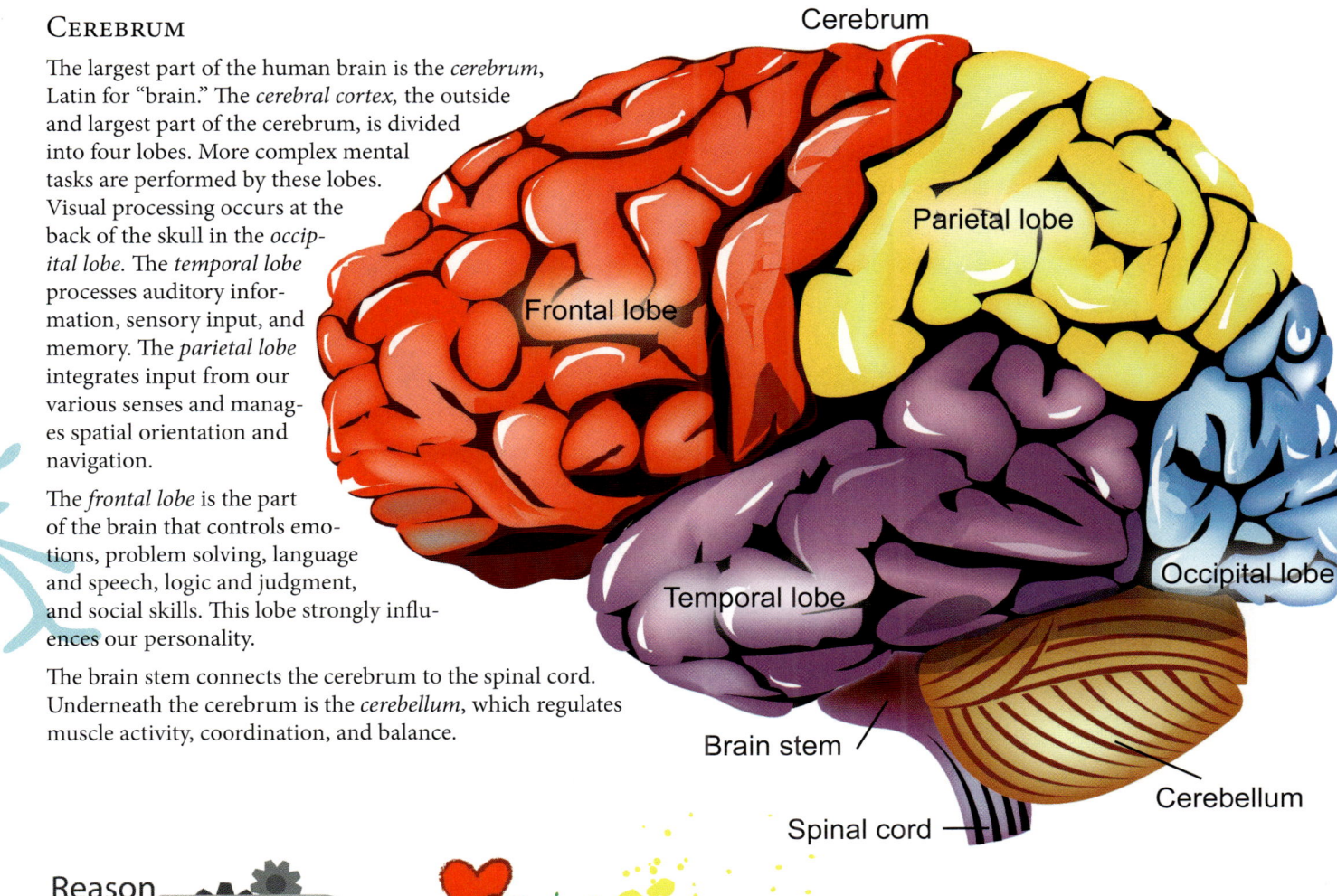

The Two Hemispheres

The human brain has two hemispheres, a left and a right. The left side of the body is mainly controlled by the right hemisphere and vice versa. One hemisphere may be slightly dominant, as with left- or right-handedness.

The left hemisphere specializes in speech and language, mathematical calculation, and analytical problem-solving. The right hemisphere, in contrast, manages visual and artistic ideas—our more creative endeavors. Though each side has its specialties, both hemispheres communicate as they perform each task.

How do the two separate halves of the brain work together? They are joined by a massive bundle of 20 million nerve fibers called the *corpus callosum.* These millions of connections enable the back-and-forth communication needed to fulfill any variety of tasks. Everyone uses both brain halves all the time since all parts are connected.

Scientists once believed that the brain loses its ability to form new neural connections when a person reaches adulthood. This powerful learning ability was thought to be confined to infancy and childhood. But a study on a stroke patient found that her brain had adapted to the damaged nerve that carries visual information by pulling similar information from other nerves. It was as if her brain had paved a new road, allowing information to get through by another route. Other studies confirm that human neurons can make new connections well into adulthood. These astounding brain capacities came as a surprise to secular researchers, who didn't expect to find such generously supplied and expertly crafted learning flexibility.

Did you know?

Our brains are energy hogs. This organ accounts for only about 2% of our body weight, but it uses about 20% of the oxygen and 25% of the glucose in our blood.

Journey from Thought to Action

A ballerina knows firsthand the elaborate coordination of multiple body parts required to accomplish athletic and artistic feats. As she waits in the wings, preparing to enter the stage, what part do you think she will engage first? Maybe she'll lift her arms as she stretches them out for balance. Or perhaps she'll rise to her toes. Both are reasonable predictions. But actually, a grand performance of any kind will always begin with the brain.

HOW DOES YOUR BRAIN TELL YOUR BODY WHAT TO DO?

The nervous system always keeps tabs on the position of your limbs, regularly scanning the location of joints and muscles like a GPS. This information is essential for any action, enabling the brain to control the body's activities.

Action starts as an idea—you decide to act and signal the process to begin. Ideas cannot be weighed or measured, and no one knows yet how they actually start. Your brain then performs three main steps to initiate a voluntary movement.

MAKE A PLAN

The brain's first step is to plan what you intend to do. For the ballerina, this may involve selecting whether she will perform a grand jeté, pirouette, or plié. This process is first detectable in the cortex, the conscious part of your brain. The parietal (sensory) and frontal (thinking) lobes are employed first.

DETERMINE THE STEPS

Next, the brain determines what sequence of muscle contractions are required to carry out the planned action.

This step requires the motor cortex (motor control) to cooperate with the cerebellum, a gatekeeper that plays multiple roles. The cerebellum stores learned sequences of movements, participates in fine tuning and coordinating movements initiated elsewhere in the brain, and integrates all of these things to produce fluid and harmonious movements.

If you've kicked a soccer ball before, your brain doesn't have to determine the steps from scratch. The supplementary and premotor cortical areas of the brain recall memories of actions you've done in the past.

DID YOU KNOW?

You build "muscle memory" as you perform repetitive tasks like throwing a baseball or playing a violin. Muscle memories are stored in your brain, not in your muscles. It takes around 10,000 hours for someone to build enough of these kinds of memories to become proficient at a complex physical task.

Just Do It!

Third, the brain executes the planned action: the ballerina pliés, your foot kicks the ball, the violinist slides the bow. Motor neurons trigger these movements, while information from subcortical structures such as the basal ganglia and the primary cortex cause the action to be carried out. The premotor cortex, motor cortex, and cerebellum work together in a continuous circuit, sending specific signals from your brain through the spinal cord and nerves to the extremities. Neurons touch specific muscle groups and tell them to fire, muscles contract, and the body moves according to plan.

Perception and Reaction

The nervous system works on the principle of input and output that we recognize as perception and reaction.

For example, when you accidentally touch a hot stove, you'll immediately jerk your hand away. Such a quick response may have happened with little time to think, but surprisingly the whole brain is involved. The moment you feel the heat, your brain signals the danger and tells the muscles of your hand, arm, and shoulder to move your hand out of the way, protecting you from a much more serious injury.

Muscle Memory

Muscle memory is stored in your brain as a collection of frequently performed muscle tasks. It's a type of repetition-formed procedural memory that can help you become very skilled at a complicated activity. "Practice makes perfect" is an accurate phrase because the more you do something, the more you build up that specific procedural memory. It's why you'll always remember how to ride a bike. You've been building muscle memory since early childhood, and your brain can tap into these memories and instruct your muscles to carry out specific tasks in a fraction of a second!

Did you know?

The average reaction time for humans is only about 150 milliseconds for a touch stimulus like heat—that's less than one-sixth of a second.

Temperature Control

Have you ever wondered why you get hot or cold? Or why you sweat after a workout or shiver when you walk in the snow? The body's adjustment involves much more than just surface changes—it goes way beyond skin-deep. In fact, when you feel sweat beading on your brow or you see goose bumps appear on your arms, your brain is hard at work. One small area of that control center of your body—the brain—is making constant modifications, keeping up with your body's need to adjust to temperature changes.

Hypothalamus

98.6º

DID YOU KNOW?

About 60% of your daily energy needs is expended just staying alive, and that's not counting energy needed to simply move around.

Every part of the brain helps regulate body temperature, but the hypothalamus does most of the work. Located behind your eyes, it determines whether your body should generate or lose heat, much like a thermostat controls the heating and cooling of your home.

Though the hypothalamus keeps the body around 98.6 degrees Fahrenheit, the "safe" temperature range for vital organs, also known as core temperature, varies from 96 to 101 degrees. Any variance of core temperature by more than 10 degrees Fahrenheit above or below this range could be deadly, as proteins—particularly enzymes—start losing both their shape and function outside the safe range.

SWEAT

The hypothalamus controls cooling of the body in a few different ways, but the most obvious one is through sweating. The middle layer of skin called the *dermis* releases tiny droplets of water that can easily evaporate, keeping your body from overheating while you're mowing the lawn or playing baseball on a summer day. In extreme heat, a person's skin can evaporate up to two quarts of sweat per hour. But in mild warmth, the body simply relaxes its muscles, allowing blood vessels to dilate and carry more blood to the limbs. Skin then acts as a giant radiator that offloads body heat to air currents that carry it away.

DID YOU KNOW?

The better physical shape you are in, the sooner you'll start sweating when you exercise. The body learns to cool off more quickly, enabling a longer workout.

HEATSTROKE

Heatstroke occurs when the body is exposed to heat for so long that it runs out of sweat, hindering the body's primary method of cooling down. If you must spend time in scorching temperatures, you can help prevent heatstroke by taking regular water breaks and resting in the shade. Symptoms include dry skin, a rapid heartbeat, throbbing headache, dizziness, and nausea. If you think you're suffering from heatstroke, be sure to cool yourself down as soon as possible. One method is to douse your body with water and sit by a fan.

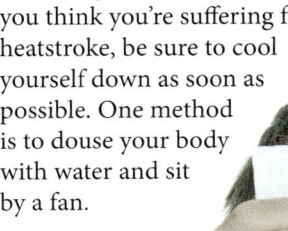

FEVER

Sickness can cause the core temperature to rise above 98.6 degrees, referred to as a fever. The body uses fever as a tool to combat harmful microbes that don't thrive in higher temperatures, and the increased heat seems to boost the immune system's performance.

HYPOTHERMIA

Prolonged exposure to cold and a drop in core temperature to 95 degrees or below can lead to hypothermia. Symptoms include shallow breathing, confusion, drowsiness, and loss of coordination. The body attempts to warm itself by shivering, which makes muscles active and produces heat. As long as you're shivering, it means your body's regulatory systems are still intact. If the core temperature becomes too low, you may stop shivering altogether—and that's not a good thing!

The Endocrine System: Second in Command

If the brain were your body's president, then the endocrine system would be the vice president. Second in command, the endocrine system serves as a control center for multiple functions of the body. Glands in the endocrine system produce hormones—chemicals that function like messenger molecules. Hormones help regulate numerous processes, including metabolism, growth, reproduction, sleep, and mood. It's normal for hormone levels to fluctuate based on environment, stress levels, gender, and age. But healthy glands will maintain them in balance, making sure everything in your body continues running smoothly.

HYPOTHALAMUS

The central nervous and endocrine systems must work together to regulate vital body functions, but they need an interface that speaks both of their "languages"—nerves' electrochemical impulses and endocrine's hormone signals. The hypothalamus translates nerve impulses to hormone production, which displays God's clever design. In healthy bodies, the hypothalamus makes its own hormones but also stimulates the pituitary gland to release just the right hormones at just the right times and lengths of time.

PITUITARY GLAND

The pituitary gland is often called the Master Gland because it manages most of the other glands that make up your endocrine system. It resides just below the hypothalamus, from which it receives signals that it transmits throughout your body. Your pituitary gland releases hormones that control growth as well as stimulate your thyroid, adrenal, and reproductive glands like the gonads—each at just the right time.

THYROID GLAND

The thyroid gland releases hormones that stimulate growth and development, and regulate body temperature and metabolism. Thyroid hormones signal bones to absorb calcium for strength and durability. The thyroid gland is crucial in developing the brain during infancy and childhood.

PARATHYROID GLANDS

The sole purpose of the parathyroid glands is to regulate the amount of calcium in your blood. They do this by producing more of a parathyroid hormone whenever your calcium is too low and producing less when it is too high. When needed, these hormones regulate the release of calcium from your bones to use in your blood, which affects the strength and density of your bones.

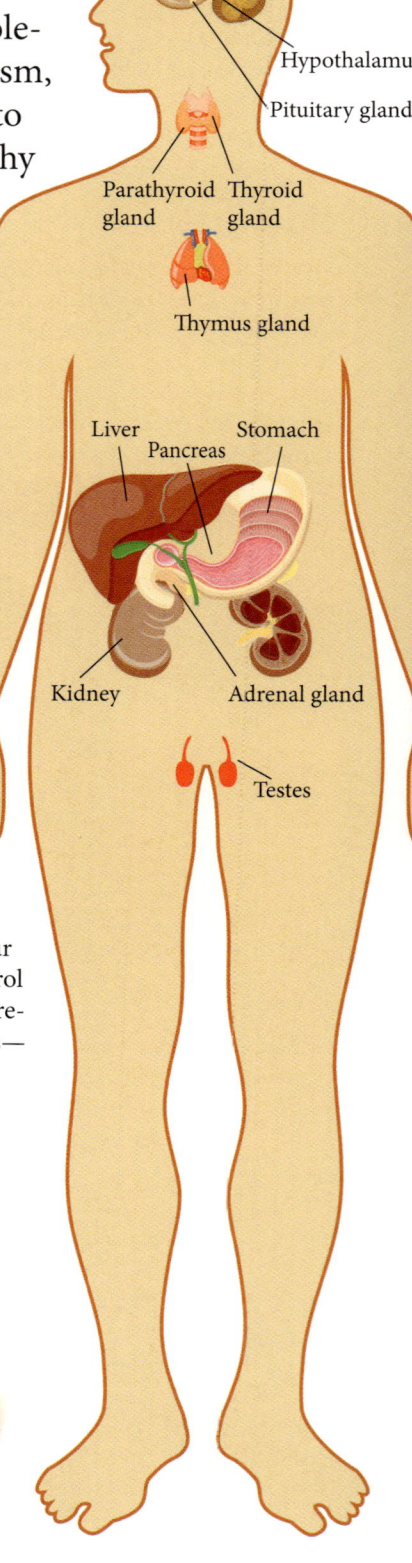

Pineal gland

Hypothalamus

Pituitary gland

Parathyroid gland　Thyroid gland

Thymus gland

Liver　Pancreas　Stomach

Kidney　Adrenal gland

Testes

PANCREAS

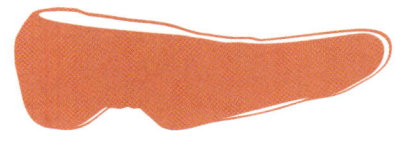

Ninety percent of the pancreas is dedicated to digestive functions, but the rest contributes to the endocrine system by using insulin and glucagon to maintain the body's blood sugar balance. Insulin allows your body's cells to absorb and use the sugar glucose. Having too much or too little glucose can cause serious health problems. Diabetes is caused by improper glucose levels, so it's closely associated with the pancreas.

PINEAL GLAND

The pineal gland provides melatonin to regulate your sleep cycle. Its release of melatonin is actually controlled by the amount of light that hits the retina of your eye. When there is little to no light, the pineal gland produces melatonin, helping you to fall asleep. But when light hits the retina, the flow of melatonin ceases. That's why it's difficult to sleep with lights on.

MALE GONADS: TESTES

Testes secrete testosterone, which grows and develops the male body for reproduction and maintains it during adulthood.

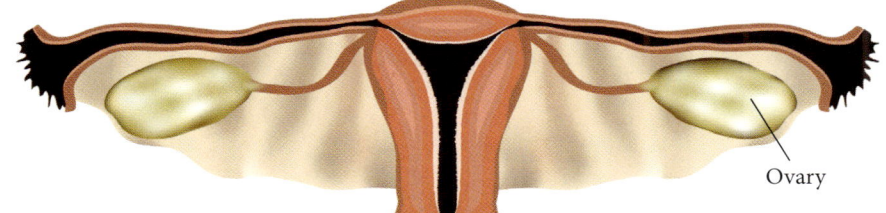

Ovary

ADRENAL GLANDS

The adrenal glands are mostly known for secreting adrenaline, the hormone that prepares your body for stress. But they also release hormones to control inflammation and immune response, the concentration of mineral ions in your body, blood pressure, and testosterone production.

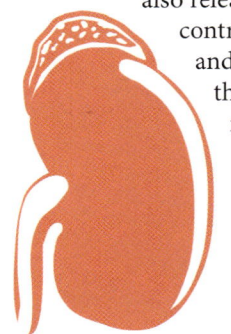

FEMALE GONADS: OVARIES

Ovaries secrete two main hormones vital to reproductive development: estrogen and progesterone. Estrogen prepares the female body for reproduction. Progesterone causes the uterine lining to thicken for pregnancy.

DID YOU KNOW?

Diabetics can measure their blood glucose levels with a glucometer. This is especially recommended for those who take insulin.

DIABETES

Diabetes is a disorder that results from the body's inability to process the sugar glucose. It is typically linked to obesity, diet, and family history, but some individuals have diabetes without links to any of these. Diabetes comes in two types. Type 1 is a lack of insulin. The body does not produce enough insulin to promote the absorption of glucose in cells, and so individuals must inject themselves with synthesized insulin. In Type 2 diabetes, an individual's pancreas may be able to produce insulin, but his body doesn't respond to it properly. Type 2 accounts for 90-95% of all diabetes disorders. Symptoms of both types include extreme fatigue and thirst, increased urination, blurry vision, and wounds that don't heal well.

The Immune System: Interfacing with the Microbial World

How does a designer get two unrelated things to work together? He must design an interface system to bridge gaps and control reciprocal actions. Our immune system is the interface between us and our microbial companions living in and on us. When microbes go where they don't belong, they face an army of cells equipped with a sophisticated array of weaponry ready to search and destroy. The immune system strategically controls microscopic organisms like bacteria, viruses, fungi, and parasites and remembers them to either work with them or defend against them if they attack the body. Without this innate, God-given control system, none of us could survive, much less reach maturity.

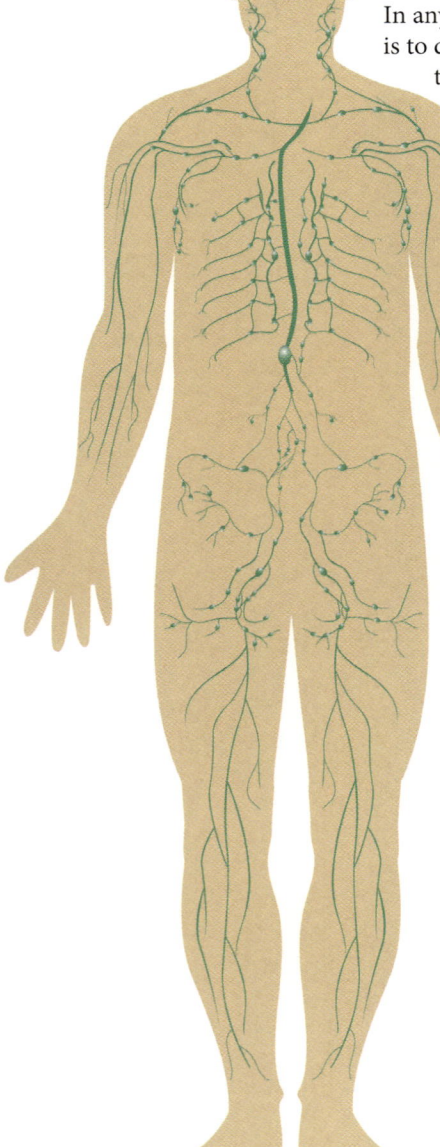

CELL MARKERS

In any relationship between two participants, the first priority is to distinguish between self and non-self. Your body does this through chemical markers. Any cell with this marker is considered "self" and should be left alone. Any cell or organism without it in a location it should not be is destroyed. The markers differ from person to person.

Antigen-presenting cells (APCs) faithfully stand guard at vulnerable openings to the body, such as the mouth, lungs, or broken skin. Microbes often attempt to enter the body through these channels. When the APCs detect a foreign invader, they engulf it, break it apart, and present portions of it called antigens to the system. This alerts the rest of the body to destroy it.

DID YOU KNOW?

After neutrophils engulf a bacterium, it is killed when a container filled with chemicals like hydrogen peroxide destroy its cell wall.

LYMPHATIC SYSTEM

The lymphatic system's ducts and valves transport loose fluid back into the circulatory system, checking it for invaders on the way.

NEUTROPHILS

Neutrophils make up about 62% of all white blood cells (WBCs). They exist only about one to two days but are produced in high quantities and replaced often. Their primary role is to attack bacteria and fungi before those materials attack your body.

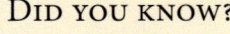

LYMPHOCYTES

Lymphocytes are divided into two categories: B-cells, which battle infections and bacteria, and T-cells that can also battle viruses and cancer cells. While lymphocytes are not as abundant as neutrophils, some last for years. They provide your immune system with a "memory" of prior exposures. That's why our response to previously encountered pathogens is more vigorous and accurate. B-cells can transform into another type of cell that makes microbe-controlling proteins called *antibodies*. These molecules can neutralize some toxins, and when they attach to a microbe, they mark it for destruction.

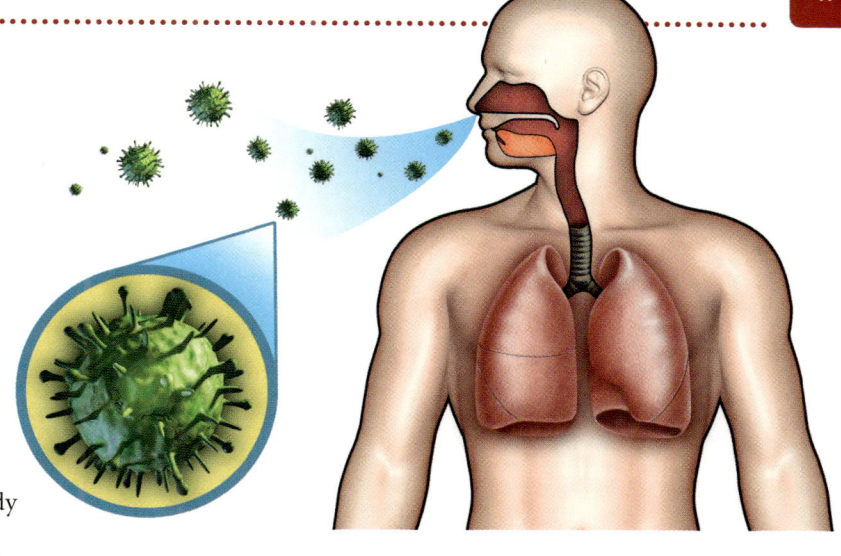

LYMPH NODES

Lymph nodes are the locations where APCs present their antigens and get a microbe-controlling response started. Lymph nodes can become enlarged and sore when your body is fighting a particularly dangerous infection. They contain lymph, a clear fluid that carries WBCs to the rest of the body. Lymph vessels collect extra fluid and recycle it back into the blood.

BONE MARROW

Bone marrow is the site of WBC production. It contains stem cells that give rise to WBCs. Stem cells are valued for research because they are able to morph into any kind of human cell.

THYMUS

The thymus is where T-cells are "programmed" to recognize self from non-self. Therefore, it is very large in newborns and children but smaller in adults. The thymus gets its name from the bud of a thyme plant because their shapes are similar.

Spleen

TONSILS

Tonsils have tissue similar to lymph nodes and are covered by pink mucosa, the same tissue that covers your tongue. They detect bacteria and viruses that enter through your mouth or nose.

SPLEEN

The spleen is the largest lymphatic organ in your body that stores WBCs. It also helps regulate the amount of blood in your body and destroys old or damaged red blood cells.

Louis Pasteur, a Bible believer, made several discoveries, including his pioneering work on the immune system with vaccination. He theorized that the body might be able to learn to defend itself from a disease if exposed to small amounts of non-lethal versions. Although he didn't know it at the time, he was testing the pathogen memory of B-cell lymphocytes. On May 5, 1881, Pasteur successfully vaccinated 24 sheep, one goat, and six cows. In 1885, he vaccinated a nine-year-old boy who had been bitten by a rabid dog. This boy is the first known human to survive rabies, which is usually fatal if left untreated.

Pursuing Good Health for a Purpose

God has blessed you with an amazing body! Now it's your job to take care of it. Eating nutritious food, exercising, and maintaining a healthy weight will help your body stay in good condition to live the life God has planned for you.

"For we are His workmanship, created in Christ Jesus for good works, which God prepared beforehand that we should walk in them." (Ephesians 2:10)

LOSING AND GAINING WEIGHT

Maintaining a healthy weight can be a struggle for many, but the principle for weight gain and loss is pretty simple. You gain weight when your body stores more energy than it uses. You lose weight when you use more energy than your body is taking in. When you lose weight, your body primarily breaks down fat. The number of fat cells doesn't decrease, each cell just gets smaller.

MEASURING BODY MASS INDEX

Body Mass Index (BMI) is a calculation comparing a person's height to their weight. BMI helps assess whether they are underweight, overweight, or at a healthy weight. While a BMI chart can be helpful, it is only a guide. You should see a health practitioner for further assessment if you do not fall within the healthy BMI range.

DID YOU KNOW?

One pound of fat equals about 3,500 calories.

Even though people strive to burn fat for weight loss, not all fat is bad. Everyone needs good sources of lipids in their diet to keep the body running properly. Some examples include avocado, eggs, olive oil, nuts, and salmon. Foods with healthy lipid content are particularly beneficial for the brain and replenishment of cells.

Were Adam and Eve Vegetarians?

"And God said, 'See, I have given you every herb that yields seed which is on the face of all the earth, and every tree whose fruit yields seed; to you it shall be for food.'" (Genesis 1:29)

When Adam and Eve were living in the Garden of Eden, God gave them fruits and vegetables to eat. It was not until after the Flood that God expanded the menu to include meat for Noah and his descendants (Genesis 9:3). Even so, public health research shows that a diet high in fruits and vegetables still provides many benefits for us.

Counting Calories

A calorie is a measurement of energy. Since food breaks down into energy, the fats, proteins, and carbohydrates found in food are measured in calories. One gram of fat contains almost nine calories of energy, while one gram of carbohydrate or protein contains four. With this knowledge, you can calculate the total number of calories in any food as long as you know how many grams of each energy source it contains.

Most experts agree you can maintain a healthy weight by eating a reasonable balance of carbohydrates, fat, and protein, monitoring your food intake, and exercising regularly. A diet of 1,500–2,000 calories per day is recommended for the average person. However, if you are living a healthy lifestyle and still find you are not at a healthy weight, abnormal levels of certain hormones might be the culprit.

Get Moving

We all know exercise is key in living a healthy lifestyle. If you can, you should try to exercise 30 minutes every day. Fast walking, jogging, or running combined with some strength training is a good combination. Exercise is also an important component in maintaining a healthy weight and lowering your risk of disease. These rewards are worth the effort in themselves, but there are bonus benefits—lifting your mood, boosting energy, and promoting better sleep.

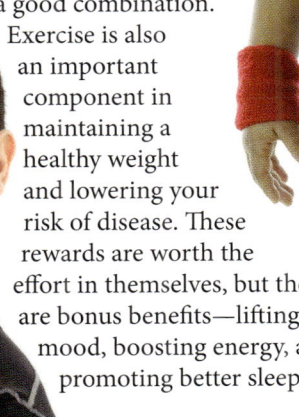

Sound Sleep and Sweet Dreams

After a day full of chatting, laughing, exploring, cooking, running, learning, serving, and more, it's time to give your body a break! Sleep enables the body to repair itself and regroup for the next day's activities. In fact, it's so important that we spend about one-third of our lives doing it!

When you sleep well, you wake up refreshed and alert for daily activities. Sleep affects how you feel, look, and perform, and it can have a major impact on your quality of life. Both the quantity and the quality of sleep are important. Sufficient sleep helps you thrive by supporting a healthy immune system, muscle repair, brain processing, and hormone regulation.

If you don't get enough rest, your body will not function efficiently and your health can break down. People who are sleep-deprived can have memory problems, a weakened immune system, and reduced hand-eye coordination.

SLEEP CYCLES

When you sleep, you move between two sleep states: the rapid eye movement (REM) and non-rapid eye movement (NREM) phases. Each full sleep cycle lasts 90-110 minutes. The average person spends about 20 minutes of each cycle in REM sleep.

As you begin to fall asleep, the NREM stage begins. This light sleep disengages you from your surroundings and your body temperature drops.

The next phase of NREM sleep is the deepest and most restorative. Your blood pressure drops, breathing slows, and muscles relax. The blood supply increases in muscles while your body works to grow and repair tissue. Energy is restored and hormones are released.

REM sleep first occurs about 90 minutes after falling asleep and recurs about every 90 minutes. Your eyes dart back and forth beneath the eyelids during this stage, giving REM its name. Heart rate, blood pressure, and breathing increase as well.

During this time, your body becomes immobile and relaxed—seemingly paralyzed. Though the body relaxes during sleep, your brain remains quite active, though it is recharging in its own way. Your sleeping brain can never fully go "off duty" because it continually controls many vital body functions like heart rate, digestion, and breathing. It is also busy dreaming, particularly during the REM period.

> ### DID YOU KNOW?
>
> A small extension of the brain called the *pineal gland* releases a nocturnal sleep aid hormone called *melatonin* to help regulate sleep cycles.

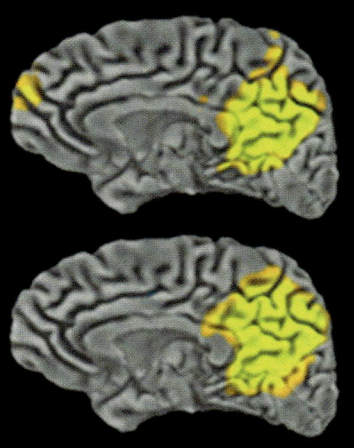

MRI images of an awake brain (top) and a brain in deep sleep (bottom).

DID YOU KNOW?

Most adults need between seven and nine hours of sleep each night to maintain optimum health and physical performance. But teens need between eight and ten hours, and infants require 14-15 hours!

WHAT IS A COMA?

A coma is an extended state of unconsciousness in which a person is unresponsive to their surroundings. The person is very much alive and appears asleep, but the person cannot be awakened by stimulation, even pain. Most comas come from some type of brain injury.

WHY DO WE DREAM?

Scientists still don't know exactly how and why we dream, but explanations have been attempted through several dream theories. Dreams may help the brain sort through the information it collects while awake. Each day you receive countless bits of sensory data—what you see, hear, taste, smell, and touch must be processed by the brain. During sleep, the brain may work to categorize and file this information, and dreams appear to play a role in this process.

Research also shows that dreams help us form memories. Dreaming allows your brain to reshuffle its memories, keeping the important ones and getting rid of the less valuable ones. Perhaps dreams help convert short-term memories into long-term memories.

Another theory is that dreams help us deal with emotions. While sleeping, your brain makes connections that it wouldn't make while awake, and that may help you understand the cause of an episode of anger or fear.

Dreams also may aid creativity. For instance, some musicians credit their dreams for helping them write a particular song. A dream may help you think in imaginative ways and answer complicated questions.

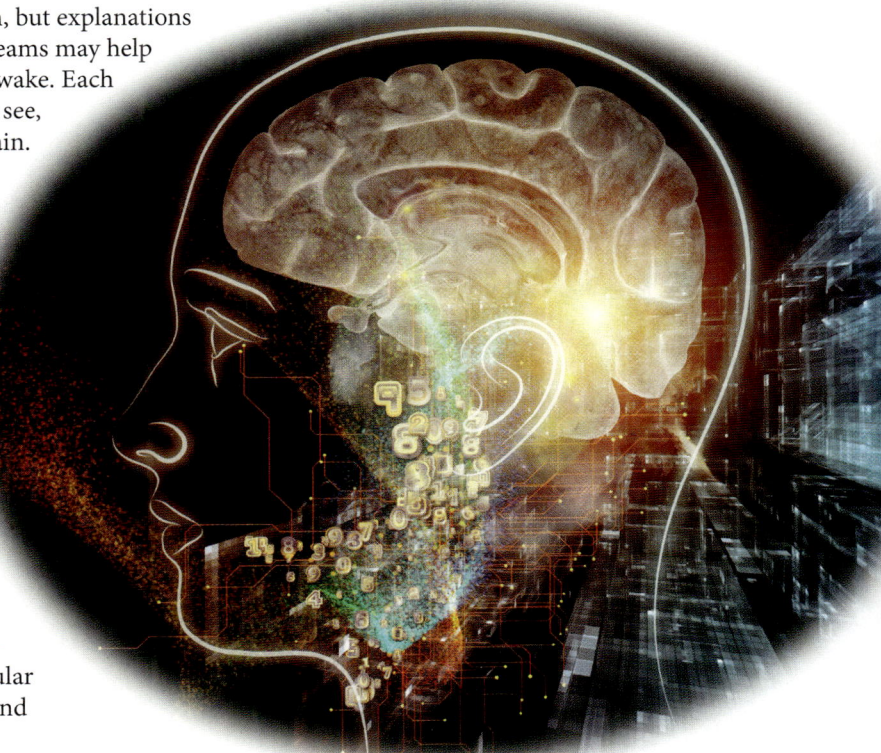

The Five Senses: Connecting You to the World

A harvest moon, contagious laughter, pumpkin pie, an encouraging hug, and backyard honeysuckle are all bound together by one common factor: we could not enjoy any of these pleasures without one of our five senses—the ability to see, hear, taste, touch, and smell. Not only do these senses allow us to delight in good things, but they also help us anticipate danger and connect with the world around us. Eyes, ears, tongue, skin, and nose all receive sensory data through strategically placed sensors and detectors that communicate data about our environment to the body's nervous system, which interprets the data into information.

Eyes Wide Open

Seeing begins by receiving light through the dark pupil in the center of your eye. The colored part of your eye, the iris, changes the shape of the pupil to help eyes adjust to different amounts of light. The lens at the front of your eye adjusts its shape to focus the light onto the retina in the back of the eye.

The retina contains numerous photoreceptors that convert the light into electrochemical signals. The two types of photoreceptors are rods, which see in dim lighting, and cones, which receive clear, well-lit images and enable you to see color. These photoreceptors initiate electrochemical impulses that communicate information to your brain. The brain then catalogues the information and matches it to past memories to help you recognize what you are seeing.

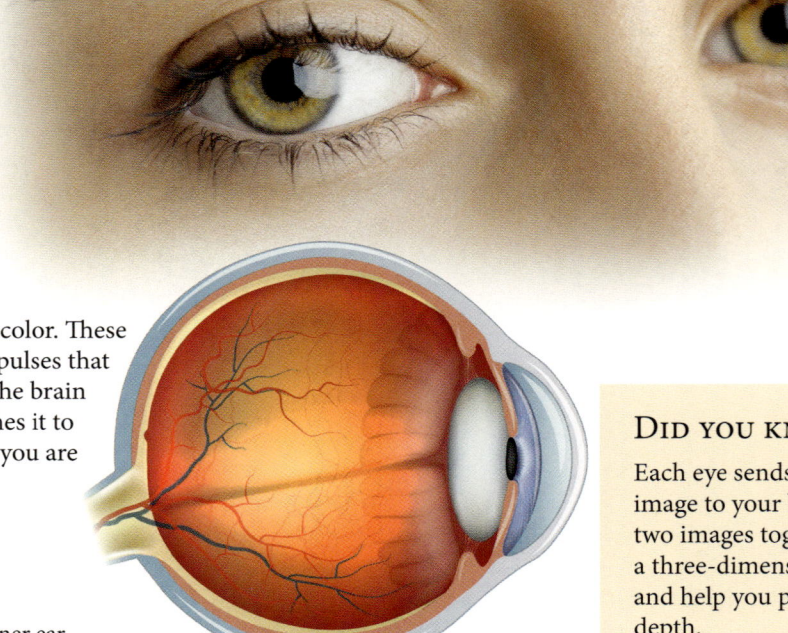

Did you know?

Each eye sends a separate image to your brain. The two images together create a three-dimensional image and help you perceive depth.

Outer ear

Inner ear

Middle ear

Cochlea

Eardrum

Ossicles

In One Ear and Out the Other

Your ear is the perfect shape for hearing since it is curved almost like a cup, designed to efficiently receive sound. The cup-like part that you can see is called the outer ear. Sound travels through the outer ear into the ear canal and strikes your eardrum, causing it to vibrate. The middle ear—the space behind the eardrum—contains three tiny bones called *ossicles*. The vibrations from the eardrum carry over to the ossicles, moving fluid around in the inner ear, the cochlea. The fluid in the cochlea moves hair cells that send electrochemical signals through the auditory nerve to the brain. The brain then recognizes the sound signal and can also determine where the sound is coming from and estimate how far away it is using the information it receives from both the left and right ears.

Right on the Nose

The sense of smell starts with your nose. The inside is lined with millions of specialized sensors. These detect odor molecules as you breathe them in and send a signal through your olfactory nerve to the brain. If it's something you have smelled before, your brain will recognize and identify it. If you smell something new, the brain must depend on other senses, such as sight or taste, to determine what the smell is.

Did you know?

Your sense of smell becomes stronger when you're hungry.

This means "I love you" in American Sign Language.

Did you know?

The taste buds you have right now will be replaced by different ones in 10-14 days. They are regularly regenerating, which is why burning your tongue on hot soup doesn't permanently damage your tasting ability.

It's on the Tip of Your Tongue

The senses of smell and taste are closely tied together. You must be able to smell something in order to taste it. And you've probably heard about taste buds—you have thousands of them—but they are not the little bumps that you see on your tongue. Taste buds are actually located within and around those little bumps, the papillae. Each bud contains between 50 and 150 sensors that respond to at least four different stimuli in your food: sweet, salty, bitter, and umami (a meaty taste). Data are sent to your brain for it to identify as "taste."

A Touching Story

Skin contains four different types of detectors interspersed all over the body to detect touch, pain, pressure, and temperature. When a detector is stimulated, an electrochemical signal is sent to the brain. The brain recognizes the sensation and determines how to react. If you did not have these sensors strategically placed on your body, you might lean your hand on a hot stove and not even know you were being burned.

DANGER! HOT SURFACE

Did you know?

If a person lacks one sense, the other senses often sharpen to compensate. A deaf person can communicate by reading lips or using sign language. Someone who is blind can read with their sense of touch thanks to braille books, which contain small bumps representing different letters of the alphabet.

A Closer Look at the Eye

Take a look around you. The gift of sight is priceless. The Lord Jesus Christ has engineered numerous parts and processes to enable your vision and provide continual care and preservation of your eyes. Key components must be in the right place, at the right time, at the right scale, and in the right amounts for you to even see the words on this page. It's a perfect example of all-or-nothing unity and precise design.

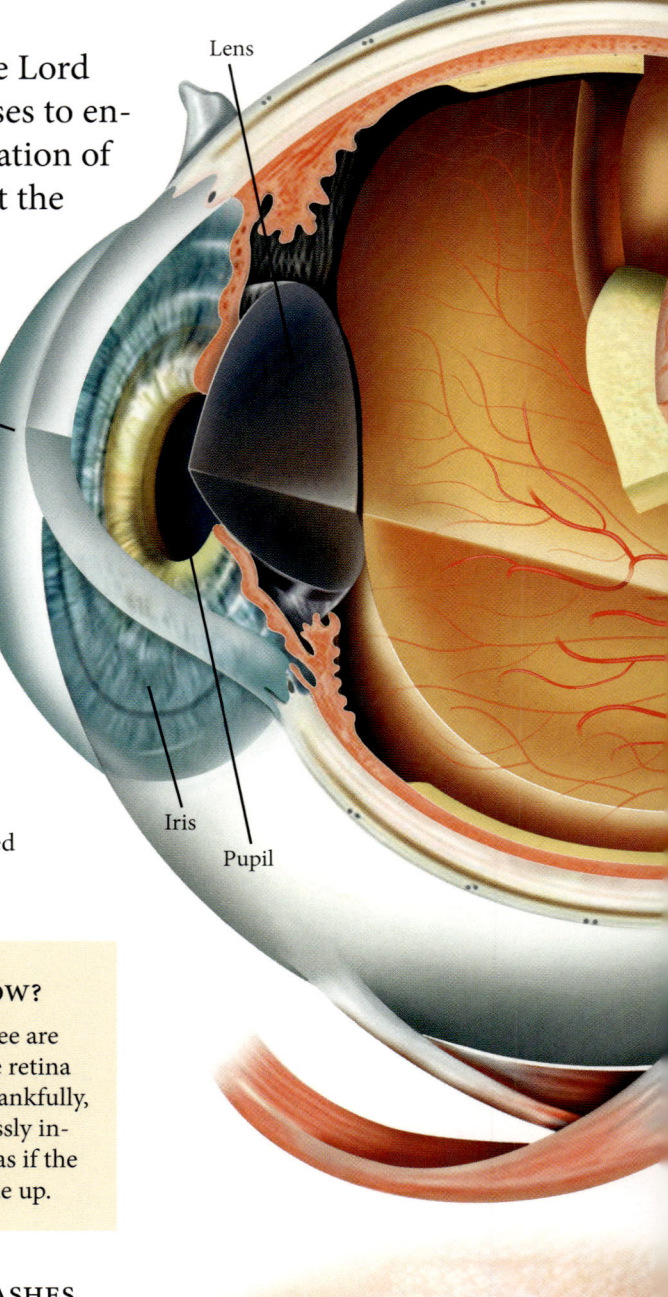

Lens

Cornea

Iris

Pupil

How Eyes Work

Sight begins when light rays reflecting off a surface enter the eye through the cornea. The cornea is the outer covering of the eye that bends light so it can enter your pupil. The colored part of the eye, the iris, adjusts the pupil's size to receive the right amount of light needed to see. Light rays then pass from the pupil to the lens of the eye, which focuses light onto the retina at the back of the eye.

Millions of photoreceptors cover the retina, ready to detect even a single photon, the smallest measurable amount of light. Rods are the photoreceptors that enable us to see in dim light, while cones let us see colors and fine detail. These photoreceptors convert the light into electrochemical impulses and send information to the brain by way of the optical nerve. The brain processes the data and associates it with catalogued memories. When organized matches are made, you get the perception of seeing.

Did you know?

The images you see are reflected onto the retina upside-down! Thankfully, the brain effortlessly interprets the data as if the image is right-side up.

Lids and Lashes

Eyelids are not just floppy curtains that hang over your eyes. The tarsal plate, positioned near the outer edge of the eyelid, allows the lid to slide tightly over the globe of the eyeball, removing unwanted debris with every blink. The lashes lining your lids help prevent debris from getting there in the first place.

Is the Human Eye Flawed?

Evolutionists often argue against the existence of a Creator by pointing out various features of design they consider to be flawed. They believe a flawed creation is an argument against an intentional and all-knowing Designer. One such feature is the retina's photo-receptors, which are aimed away from the light they detect. But an inverted retina keeps the high-powered photoreceptors close to the choroid, their source of nutrition and oxygen, and protects them from harmful radiation behind a layer of nerve cells.

Evolutionists also point to the way the position of the retina makes nerve fibers have to thread through it to connect to the brain. This entry through the retina causes a small blind spot in the vision of each eye. But the presence of a second eye renders the blind spot irrelevant. The joint images from two eyes fully compensate for it because each eye can see the small area the other eye can't. That's one reason why you don't even notice it. The brain also uses a complex algorithm to fill in the missing data, so people with only one eye don't notice their blind spot either.

Retina

Optical nerve

Meibum: A Natural Lubricant

Meibomian glands within the tarsal plates man-ufacture a special oil called *meibum* and release it through tiny ducts right at the lid margins. When you blink, tiny meibum oil droplets roll under the lid and are uniformly spread over the eye by the precise shape of the lid margin. Meibum's soft, waxy properties make the watertight seal of eyelids over our eyes possible with only mild lid pressure.

Emotional Tears

Only humans shed emotional tears. In times of severe emotional stress, molecules called *enkephalins* are released in tears. When absorbed into the body, the crying person may feel a sense of relief or mild euphoria afterward. An evolutionary story for the origin of these tears can be con-cocted, but a better explanation is that they are a sweet provision from a loving heavenly Father.

Tears for Moisture and Cleansing

Tears produced by the lacrimal gland flow down and across the eye toward drainage ducts—called *puncta*—placed on the nasal side of the upper and lower eyelids. This keeps the eye clean and moist, providing mixtures of antibodies and enzymes to destroy microbes. Generally, tear production and removal are balanced at just the right rate—they keep the eyes from drying out without flooding your face.

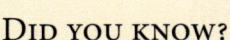

Did you know?

You get "red eye" in photos when the camera's flash re-flects off the retina of your eyes. The nearby blood vessels of the choroid are what give your pupils that seemingly red color.

Exquisite Design of Hands

And let the beauty of the LORD our God be upon us, and establish the work of our hands for us; yes, establish the work of our hands. (Psalm 90:17)

We would be hard-pressed to find a tool more versatile than the human hand. It is exquisitely designed to hold, soothe, grip, wield, pinch, tuck, strike, tap, clap, guide, beckon, lift, rub, fold, clasp, grasp, signal, squeeze, and the list goes on. Human hands possess superior capabilities that are fundamentally distinct from those of any other creature, thanks to their unique muscle configuration and the brain's extensive sensory-motor functions.

HAND BONES

Each hand contains 27 bones, granting amazing range and flexibility. Two rows of four bones in the wrists are the *carpals*. Five metacarpals serve as the framework for the palm. Each finger has three phalanges, while the thumb has only two.

DID YOU KNOW?

Scientists tried to replicate the human hand using robotics, but their best efforts only replicated about 10% of its functionality.

HAND MUSCLES

About 32 muscles control your hand. Nineteen muscles originate in the palm and are called *intrinsic*, while the other 13 muscles originate in your forearm and are called *extrinsic*. Both intrinsic and extrinsic muscles help control the four fingers and thumb. Seven muscles control just the index finger, five major muscles are unique to the thumb, and three muscles control the pinky finger.

SKIN

Your palms are constructed in a way that makes it easy for you to hold and grip things. The skin is hairless and cannot tan, making it tough and reliable. A layer of fibrous tissue connects the skin to the skeleton in a unique way that allows you to grip things without the skin moving and sliding. The creases allow your skin to fold without bunching up and getting in the way.

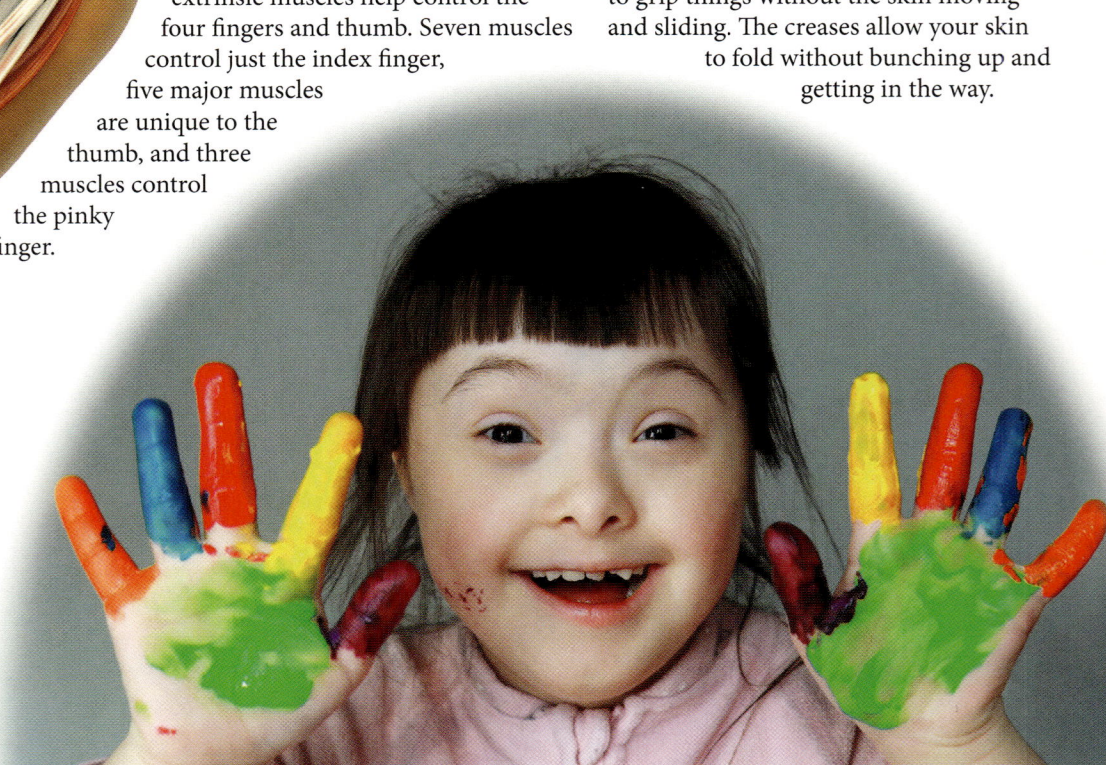

OPPOSITION

Opposition, the ability to squeeze between fingers and thumb or palm and fingers, is the most important element of hand movement. You couldn't do much if your hands were unable to grasp things. The many different kinds of muscles in your hands, combined with the brain capacity devoted to hand movement, mean that grip control combinations are infinite and remarkably versatile. Your hands can tightly hold something heavy like a hammer while grasping a fragile potato chip between two fingers of the same hand. An average man's grip has a respectable 100 pounds of force.

FINE MOTOR SKILLS

In addition to the ability to grasp things, your hands have extremely fine motor skills—especially your thumbs. Holding a pencil, tying your shoes, and other small and precise movements are made possible by a forearm muscle called the *flexor pollicis longus* (FPL). The FPL is able to move your thumb toward your palm while its tendon independently bends your thumb's tip. The FPL is not present in chimpanzees, gorillas, orangutans, or monkeys.

Your brain has such precise control over the FPL that it can activate a force of only 7/100 of an ounce! When you pinch, the brain regulates the muscles' force, but the muscles also act as sensors to let the brain know when to stop.

A HAND WITH A PLAN

Mathematical models show that the brain actually predicts fingertip actions in only 60 milliseconds, about the same time it takes to blink an eye. To obtain the highest possible finger speeds, sensors and conscious thought are augmented in the brain with an anticipatory function for individual finger movements called a *forward plan*.

Evidence shows that the central nervous system predicts the best outcome of every finger movement several movements ahead of its current state. Skilled typists will visually process up to eight characters in advance, and then the forward plan for muscle movements will commit the finger muscles to an action about three characters in advance of actually striking the keys. Time intervals between keystrokes are commonly as low as 60 milliseconds. Interestingly, speed is fastest if successive keystrokes are made by fingers on opposite hands.

These Feet Were Made for Walking

With 26 bones, 33 joints, 42 muscles, and at least 50 ligaments and tendons each, your feet demonstrate engineered features that are just as amazing as any other part of the body! They are specially designed to meet the demands of your day, whether it involves walking, jogging, running, dancing, balancing on a tightrope, or rock climbing.

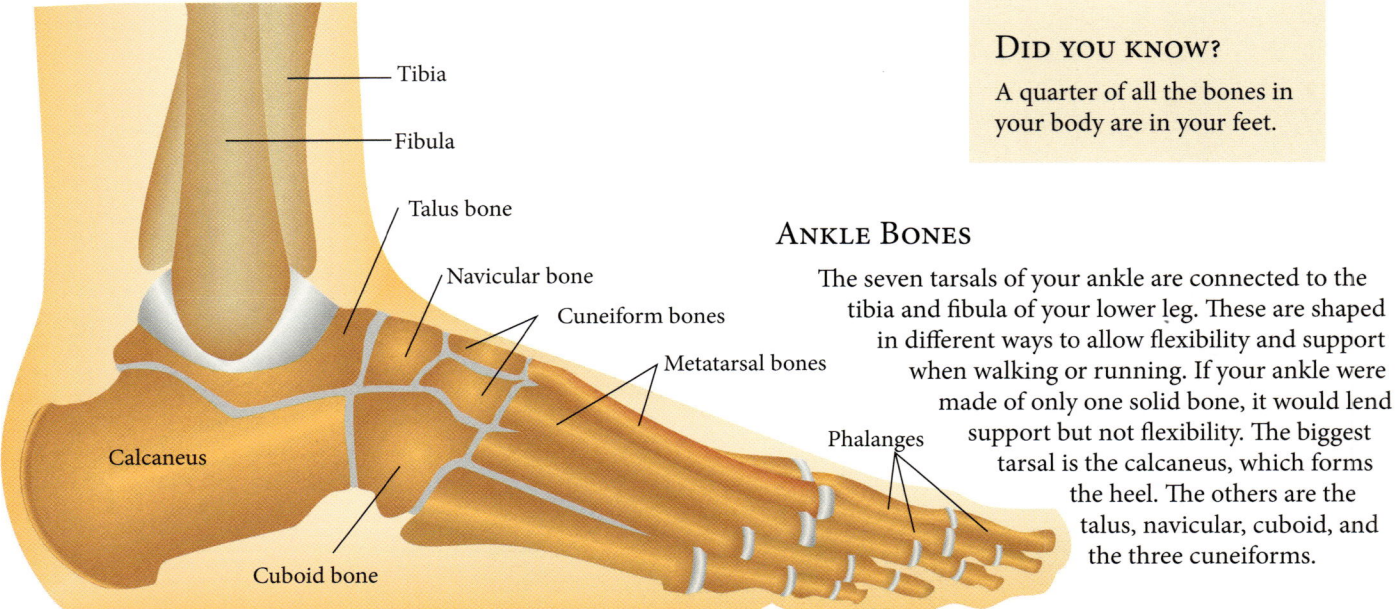

Tibia

Fibula

Talus bone

Navicular bone

Cuneiform bones

Metatarsal bones

Calcaneus

Phalanges

Cuboid bone

DID YOU KNOW?

A quarter of all the bones in your body are in your feet.

ANKLE BONES

The seven tarsals of your ankle are connected to the tibia and fibula of your lower leg. These are shaped in different ways to allow flexibility and support when walking or running. If your ankle were made of only one solid bone, it would lend support but not flexibility. The biggest tarsal is the calcaneus, which forms the heel. The others are the talus, navicular, cuboid, and the three cuneiforms.

DID YOU KNOW?

The average person takes about 8,000-10,000 steps a day. This adds up to about 115,000 miles over a lifetime. That's enough to go around the circumference of the earth four times.

MUSCLES

Most of the muscles and tendons that move the foot are located in your lower leg. They point your foot up, down, left, and right, and tilt it back and forth. The plantar flexors form the calf, and they, along with the tibialis posterior (which merges with your Achilles tendon), allow you to point your foot down. The dorsal flexors are placed on the front of your tibia and point your foot up. The fibularis longus tendon helps point your foot up as well, but it also helps swivel your foot from side to side.

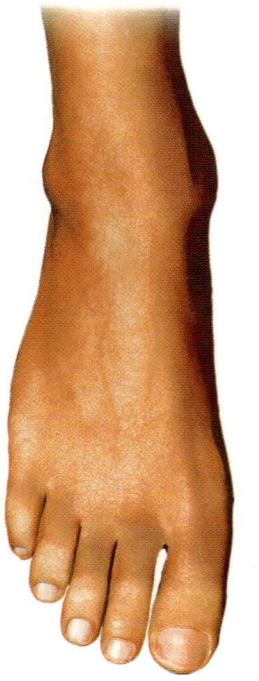
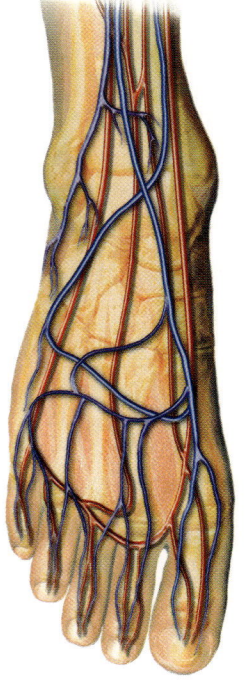
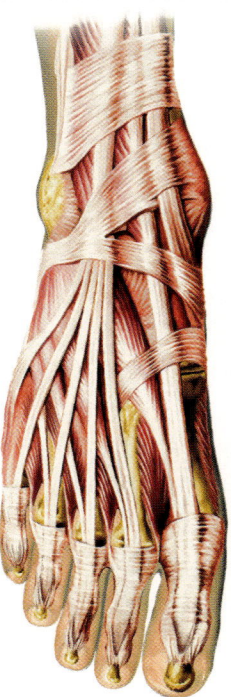
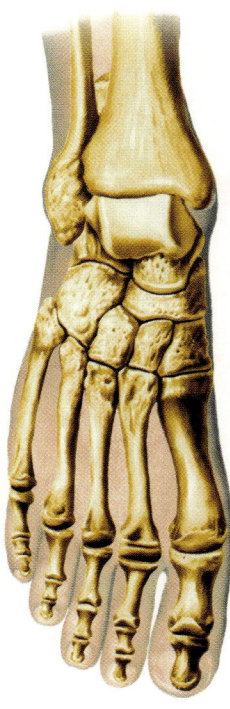

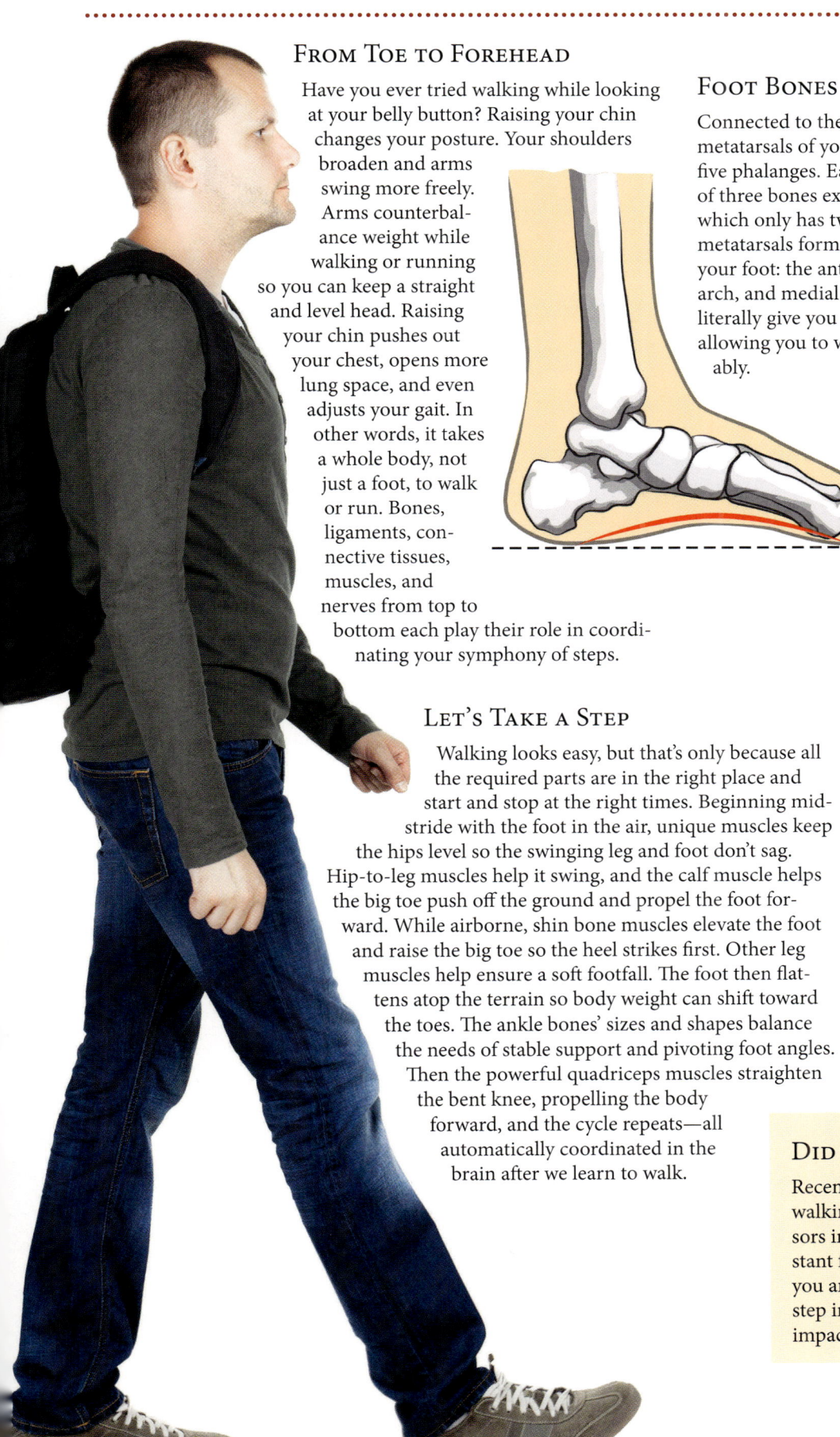

From Toe to Forehead

Have you ever tried walking while looking at your belly button? Raising your chin changes your posture. Your shoulders broaden and arms swing more freely. Arms counterbalance weight while walking or running so you can keep a straight and level head. Raising your chin pushes out your chest, opens more lung space, and even adjusts your gait. In other words, it takes a whole body, not just a foot, to walk or run. Bones, ligaments, connective tissues, muscles, and nerves from top to bottom each play their role in coordinating your symphony of steps.

Let's Take a Step

Walking looks easy, but that's only because all the required parts are in the right place and start and stop at the right times. Beginning midstride with the foot in the air, unique muscles keep the hips level so the swinging leg and foot don't sag. Hip-to-leg muscles help it swing, and the calf muscle helps the big toe push off the ground and propel the foot forward. While airborne, shin bone muscles elevate the foot and raise the big toe so the heel strikes first. Other leg muscles help ensure a soft footfall. The foot then flattens atop the terrain so body weight can shift toward the toes. The ankle bones' sizes and shapes balance the needs of stable support and pivoting foot angles. Then the powerful quadriceps muscles straighten the bent knee, propelling the body forward, and the cycle repeats—all automatically coordinated in the brain after we learn to walk.

Foot Bones

Connected to the ankle bones are the five metatarsals of your foot, one for each of the five phalanges. Each phalange is made of three bones except for the big toe, which only has two. The tarsals and metatarsals form the three arches of your foot: the anterior arch, lateral arch, and medial arch. These arches literally give you spring in your step, allowing you to walk more comfortably.

Bipedalism

Humans walk on only two legs—commonly referred to as bipedalism. It offers many advantages, such as allowing us to hold our heads higher, giving us a greater field of vision. Evolutionists have a hard time figuring out how bipedalism might have evolved. They have argued for around 100 years about when, where, and how something like that could have happened. Monkeys and others can "walk" or waddle on two feet. However, humans use bipedalism constantly, which is unique among biological creatures.

Did you know?

Recent research shows many benefits to walking barefoot. Numerous nerve sensors in the sole of your foot provide constant feedback about the kind of terrain you are walking on and how you should step in order to give the least negative impact to your joints and feet.

Cells: The Body's Building Blocks

Have you ever seen mosaic art? Remember how all those tiny stone tiles came together to compose a beautiful picture? This is the nature of the cell. We can't see an individual cell with the naked eye, but similar to a mosaic stone tile, it is the smallest, most basic building block of every organism we see. And don't be fooled by its size—if you study it closely, you'll see God has truly made the cell a work of art.

Composed of numerous interworking parts, the cell is a great example of all-or-nothing unity, meaning that it can't work if it's missing a single vital component. Evolution is based on the gradual modification of life and processes over time, and many origin-of-life theories start with the existence of a cell. The fact that the cell had to have all key parts from the very beginning in order to function gives witness to the existence of an intelligent Creator.

ABIOGENESIS

Some believe that the first cell arose spontaneously from a warm pond of primeval chemicals, a process known as *abiogenesis*. Various theories have been proposed in an attempt to explain how this mysterious process might have happened. So far none have explained the origin of the information in the cell's genome or even the proposed minimal set of 256 genes required for basic cellular life. Observations confirm that life only comes from pre-existing life, which originally began with God, who is eternal life.

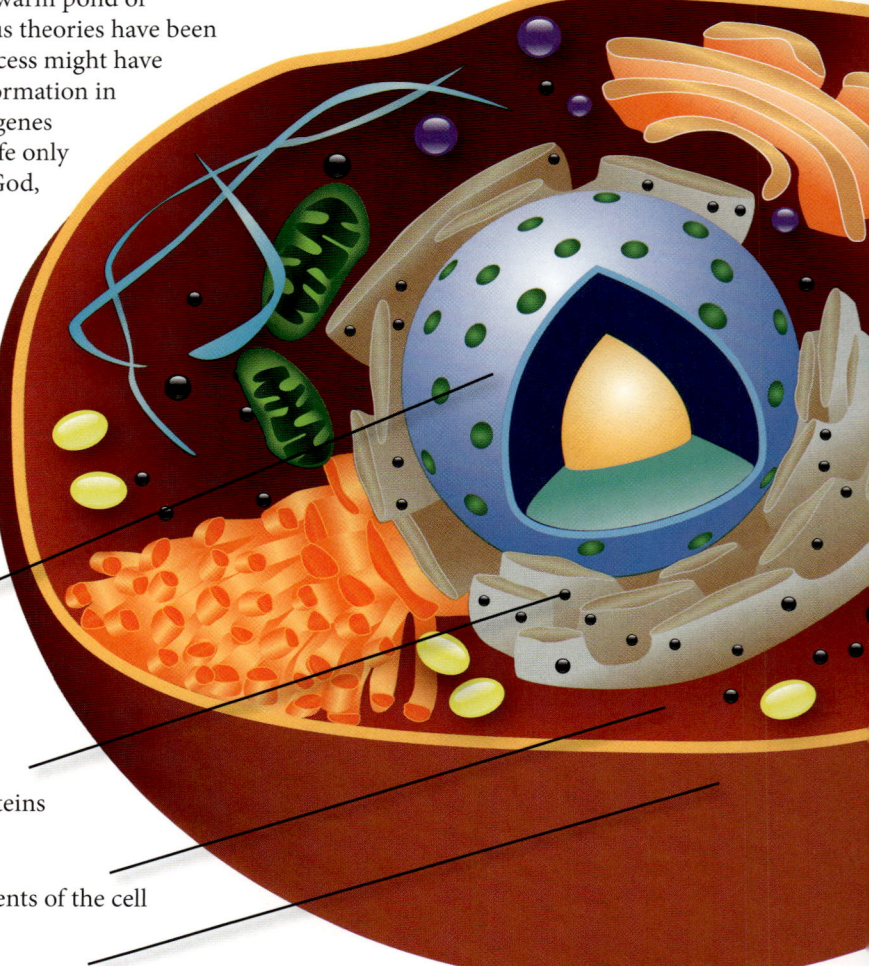

TYPES OF CELLS

Cells can be divided into two basic types. Prokaryotic cells are small, and DNA floats freely inside them. Many bacteria are prokaryotes. Eukaryotic cells are larger, more complex, and house their DNA in a nucleus. The tissues in your body contain eukaryotic cells.

NUCLEUS

Internal membrane that houses the cell's DNA (only present in eukaryotes)

RIBOSOME

Molecular machine that builds amino acids into proteins

CYTOPLASM

A semifluid substance that holds all the components of the cell

PLASMA MEMBRANE

Outer barrier that holds the cell together

ROBERT HOOK

The fact that biological life is made up of cells was first discovered in 1665 by an English scientist named Robert Hook. He was studying plant tissue under a microscope when he noticed small compartmentalized structures that looked like tiny rooms—hence the term "cell."

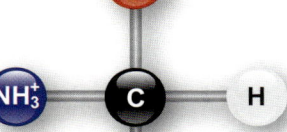

CELL DIVISION

It may sound strange, but cells actually multiply by dividing. Each cell is like a little factory, replicating its own DNA and then dividing itself into two individual cells. This process can easily be summarized as grow, divide, repeat. These images depict cell division.

DNA, AMINO ACIDS, AND PROTEINS

In the same way that cells are the building blocks of living organisms, proteins are the building blocks of cells and amino acids are the building blocks of proteins. Located inside the cell, DNA carries the coded instructions that specify which amino acids will build each protein. DNA codes mostly regulate the timing and duration of cellular activities.

DID YOU KNOW?

From the moment you were conceived, cells have been at work for your good, ensuring development in your mother's womb, growth after you were born, and renewal of your body throughout life.

DEOXYRIBONUCLEIC ACID (DNA)

Coded instructions of amino acids that direct the operations of the cell

STAGES OF CELL DIVISION

Interphase: The cell is operating normally.

Prophase: The cell replicates its DNA so that there are now two sets of chromosomes.

Metaphase: The chromosomes line up in the center of the cell.

Anaphase: The cell pulls the two sets of chromosomes apart and begins to elongate.

Telophase: Nuclear membranes form around the two chromosome sets, forming two nuclei.

Cytokinesis: The cell completely divides.

DNA: Deciphering the Code of Life

You may be familiar with the concept of DNA (deoxyribonucleic acid) since it is used for evidence in court cases or to determine if two people are related. But what purpose does it serve within the human body? DNA can be compared with the code a computer programmer writes to create a website. The programmer uses language that a computer can read and execute to create each feature of the site—color, font, layout, and so on. In a similar way, every human cell except for red blood cells contains DNA to hold biological coding instructions. DNA serves as the coding instructions, telling every cell in your body what, when, and how long to make certain products like proteins. It helps determine height, eye color, aspects of your health, organ function, facial features, physical build, and even influences your personality. Your DNA is unique from anyone else who ever lived.

Chromosome

DID YOU KNOW?

DNA is copied every time a cell divides. You have trillions of cells, and millions are dividing and making copies of your DNA right now.

Cell

DNA

DNA BASICS

Strands of nuclear DNA are quite long—about six feet long if you unrolled one cell's entire DNA! During cell division, tiny machines package DNA into chromosomes to keep it organized enough to ensure that each new cell receives a complete set of coding instructions.

We each have 23 pairs of chromosomes, 46 total, and we received half from our mother and half from our father. Humans have between 22,000 and 28,000 protein-coding genes in their genome. However, there are over twice as many other genes that make functional RNAs and do not code for proteins.

DNA is made up of four molecules or nucleotides called *adenine* (A), *thymine* (T), *guanine* (G), and *cytosine* (C). The sequential arrangement of these bases determines the genetic code. These bases occupy positions along a single DNA strand and form pairs across strands. Adenine bonds with thymine, and cytosine pairs with guanine.

The complete DNA instruction book, or genome, for a human contains about three billion nucleotide bases. Scientists use the term "double-helix" to describe DNA's twisted ladder-like structure.

THE TRANSCRIPTION PROCESS

Like copying a computer program from a hard drive, RNA polymerase copies DNA sequence from the genome onto an RNA transcript. Transcription begins when regulatory, gene-activating proteins called *transcription factors* (TFs) bind to specific sequences in and around a gene. TFs recruit many different proteins in a highly regulated process that ensures that a specific DNA sequence is copied at the right speed, the right time, and in the right amounts. Special enzymes are also required to unzip, unwind, and bend the DNA into the correct three-dimensional structure so that transcription can occur. The new transcripts, called *messenger RNAs*, are either used by the cell directly or used as templates to make proteins outside the cell nucleus at machinery called ribosomes.

GENETIC MUTATION

A genetic mutation is an alteration in a person's DNA sequence. Some mutations affect a single DNA base pair and some affect entire regions of DNA. Mutations are almost always harmful.

Germline mutations are inherited. They were either passed down from a previous generation or they occurred in the parent's reproductive tissues that formed the egg or sperm cell. In either case, the resulting child will have the mutation in all of his or her cells and could pass it on to offspring. Somatic cell mutations, like those that happen in lung or skin cells, do not occur in reproductive (germline) tissues and therefore are not heritable. However, they can lead to serious health problems like cancer.

Mutations can be caused by toxic environmental exposures or from an error in DNA replication when it gets copied during cell division. Many ingenious mechanisms exist to find and correct nearly all of the errors that occur during cell division and regular cell activity. But on rare occasions, some mutations slip through the error-checking systems of the cell.

NOT OUT OF AFRICA

Conventional evolutionary thinking places the origin of humankind in Africa about 100,000 years ago. But recent research comparing DNA sequence similarities between people groups better fits mankind originating from the Middle East less than 10,000 years ago.

Be Fruitful and Multiply

In the book of Genesis, God issues the command several times for His creatures to "be fruitful and multiply and fill the earth." God designed the reproductive system in the human body to enable mankind to fulfill this special commission.

FERTILIZATION

A new person begins the moment a human sperm cell unites with a human egg. This uniting event is called *fertilization*. Sperm cells from the father carry his contribution of genetic information on 23 of the 46 chromosomes necessary to start a new, totally unique person. The other 23 chromosomes of genetic information are inside the egg of the mother. How does this genetic information unite to form a new person? The process is far more complex and intricate than you might think, as the voyage of a single sperm cell from production to fertilization is full of obstacles that must be overcome. Were it not for the special features God designed in the reproductive systems of the male and female body to aid and encourage fertilization, no baby could ever be conceived, and the existence of humanity would have ended with Adam and Eve.

"Then God blessed them, and God said to them, 'Be fruitful and multiply; fill the earth and subdue it.'" (Genesis 1:28)

SPERM MEETS EGG

Though millions of sperm cells may initially swim toward the egg, normally less than 500 will make it to the egg, and only one sperm will achieve fertilization. When the first sperm completes its journey, its membrane unites with the egg. Then, tube-like structures rapidly form from within the egg and pull the nucleus of the sperm inside—establishing the first cell of a new person. This one cell contains a blueprint for a person who is unlike anyone else.

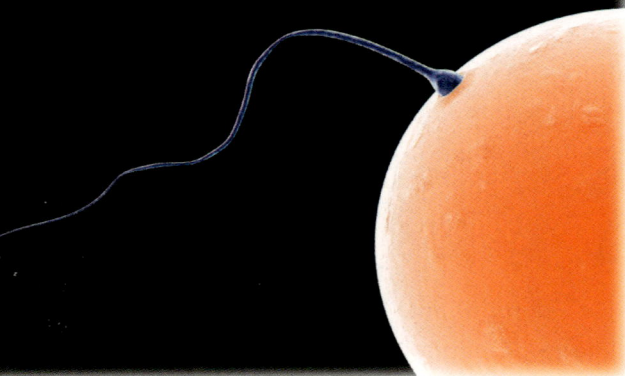

EARLY DEVELOPMENT

The new fertilized cell typically has 46 chromosomes, and the information within them determines many traits—like hair color, eye color, skin tone, height, and more. As the single cell divides and replicates itself into numerous different types of cells to develop into a baby, each cell (except red blood cells) will contain a copy of these 46 chromosomes. Within eight to nine days after fertilization, the small mass of dividing cells will implant inside the uterus of his mother, where he will develop for nine months. Mom will provide nourishment to her developing baby through a temporary interface organ called the *placenta* which is made of tissue from both the mother and baby.

DID YOU KNOW?

The first time an entire human and chimpanzee chromosome were compared was in 2010, and it was the male Y chromosome. They were remarkably divergent in structure and genes, with a 53% difference in specific gene content, a 33% difference in gene categories, and the human Y having a third more entirely different classes of genes. This finding surprised evolutionists since they often believe the mistaken notion that humans and chimps are 98-99% genetically similar.

DID YOU KNOW?

People with Down syndrome have an extra chromosome from the moment of fertilization. They share some distinctive traits, but each one has his or her own unique personality.

BOY OR GIRL?

Whether the baby will be a boy or a girl is also decided at fertilization. The egg carries an X chromosome, and the sperm carries either an X or a Y chromosome.

X+X means the new cell will develop into a baby girl. X+Y means the cell will develop into a baby boy.

DID YOU KNOW?

Ninety-nine percent of identical twins are formed when a single fertilized egg develops for one to nine days and then divides into two. They look almost exactly alike and possess genes that are nearly identical. Fraternal twins develop when two separate eggs are fertilized by two separate sperm. While they may resemble each other, their genes and gene expression patterns have subtle differences.

PERFECT HARMONY

Though this explanation of fertilization reveals some of the marvels of God's design for reproduction, it doesn't convey all of the elaborate and fascinating details. Evolutionary literature is filled with stories that attempt to explain the origin of these processes without any tested scientific evidence to support them. A more in-depth study reveals that human reproduction extends beyond the reproductive systems to the neurologic, hormonal, and circulatory systems, and involves perfect harmony between the male, female, and baby's physiologies. For such complexity there is only one reasonable explanation. These processes were placed by the Lord Jesus into the first parents, Adam and Eve, fully functional right from the beginning.

Pregnancy: A Big Job for Mom and Baby

Pregnancy turns a mother's body into a temporary home ideally suited to fit a baby's development needs. God equips a woman's body to fulfill this role, though the process often comes with significant physical and emotional challenges. Pregnancy typically lasts about 40 weeks and is divided into three stages called *trimesters*.

We sometimes view a developing baby inside the mother's body as a passive object—as if the mother's body is doing all the work required to "build" a baby while the baby is simply existing in a comfortable cocoon. Nothing could be further from the truth. From guiding implantation in the uterus all the way to breast-feeding, the baby-placenta unit is in charge of all aspects of the baby's development.

The placenta is a temporary organ that develops in the uterus to carry oxygen and nutrients to the baby. Without its help, the mother's immune system would mistake the baby for a foreign invader and destroy the new baby's first cells within just a few cell divisions. But substances secreted by the placenta and the baby suppress the mom's immune system, preventing her body from attacking. Placental tissue has fewer features that would provoke an immune response; therefore, the mother's body accepts it. Without this immunological acceptance, no baby would ever survive. If the suppression of the mother's immune system were not localized, her health could be compromised. The maternal immune system helps control implantation of the embryo at just the right depth into the uterus. Without this exact balance of immune responses, the developing placenta could invade tissue all the way through the uterine wall and be fatal to the mother.

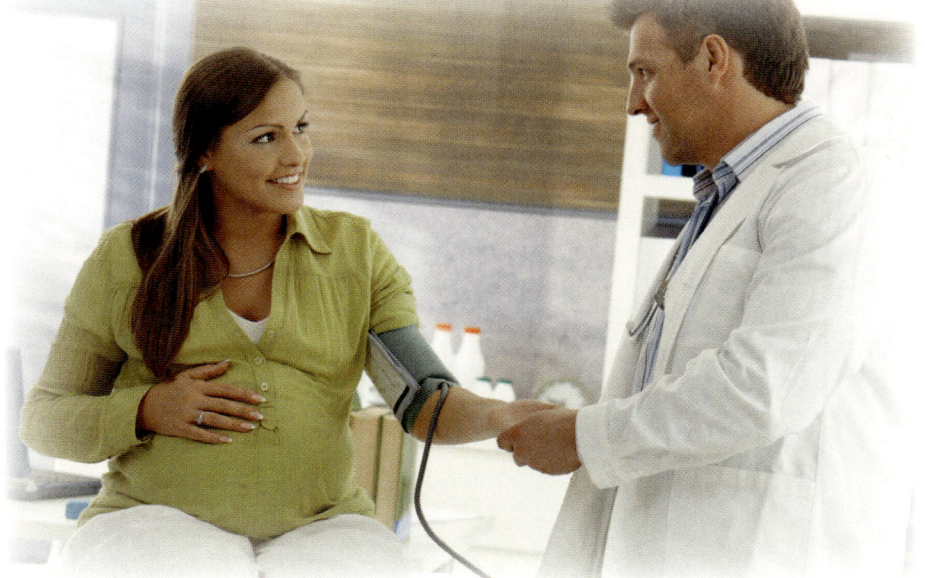

In the last weeks of pregnancy, the baby and placenta release even more hormones to signal the mom's body he is ready to be born, causing uterine contractions, appropriate relaxation of her pelvic ligaments, and the opening of the cervix, providing a birth passage for the baby to enter the world!

The baby produces hormones that stimulate the mother's body to adjust itself in many ways that are absolutely necessary for the baby's healthy development. These changes include significant increases of the mother's blood volume, cardiac output, blood pressure, blood flow to the kidneys, and metabolism. The placenta also takes nutrients from the mom's bloodstream so efficiently that the baby's needs are met first, to ensure his proper development.

While in the womb, the baby and placenta release hormones that help prepare the mother's body to begin producing milk. After delivery, newborn suckling stimulates a hormonal release by the mother's brain that lets her milk flow.

DID YOU KNOW?

After a baby is born, the nutrient content of breast milk automatically adjusts to fit precisely what is needed for each stage of the baby's development, and the mother's body even "knows" to provide extra nourishment when a baby is born premature.

The interdependence between dozens of precisely interacting parts between mother and baby is pretty compelling evidence against a step-by-step evolutionary process leading to such complex systems. Remove just one piece, one step, and the baby or mother cannot survive. Clearly, these magnificent systems were placed by the Lord Jesus in the first mother, Eve, fully functional from the very beginning.

Life in the Womb: Fearfully and Wonderfully Made

Perhaps no greater wonder exists in the human body than the processes involved in the conception and development of a baby—one human life growing within another human life. Mother and baby are deeply connected yet two very separate beings. Bible passages such as Psalm 139 tell us that God knows us intimately and has a plan for our lives even while we are still in our mother's womb.

> "Your eyes saw my substance, being yet unformed. And in Your book they all were written, the days fashioned for me, when as yet there were none of them." (Psalm 139:16)

The development of a baby is a marvelous process of construction and could not possibly happen without intentional design. Many working parts and processes must happen in order to ensure a baby's survival before and after birth. God put special features in place so that we could not only survive, but thrive.

Month 1

The fertilized egg implants in the uterus and will grow 10,000 times larger by the end of the month via cell division. At this stage, the baby is referred to as an embryo and begins forming the heart, digestive system, backbone, and spinal cord. A temporary organ called a *placenta* develops in the uterus to carry oxygen and nutrients to the baby through the umbilical cord.

Month 2

The baby's heart is now beating. During this month her eyes, nose, lips, tongue, ears, and teeth are forming. She is active, though mom cannot yet feel her tiny movements.

Did you know?

Babies form their own unique set of fingerprints only 14 weeks from conception.

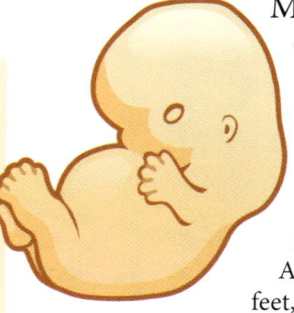

Month 3

The heartbeat can now be heard with a special instrument called a fetal Doppler. A lot of visible development occurs as her body takes on a more recognizably human form. Arms, hands, fingers, legs, feet, toes, and even earlobes finish their formation, while a little more work remains on her nails, eyes, and some organs and tissues.

Month 4

Though gender was determined at conception, a sonogram can now reveal it, and the baby may be found sucking his or her thumb. The skin is bright pink and covered in *lanugo*, a temporary downy coating of hair. Additional tasks this month are developing tooth buds and sweat glands.

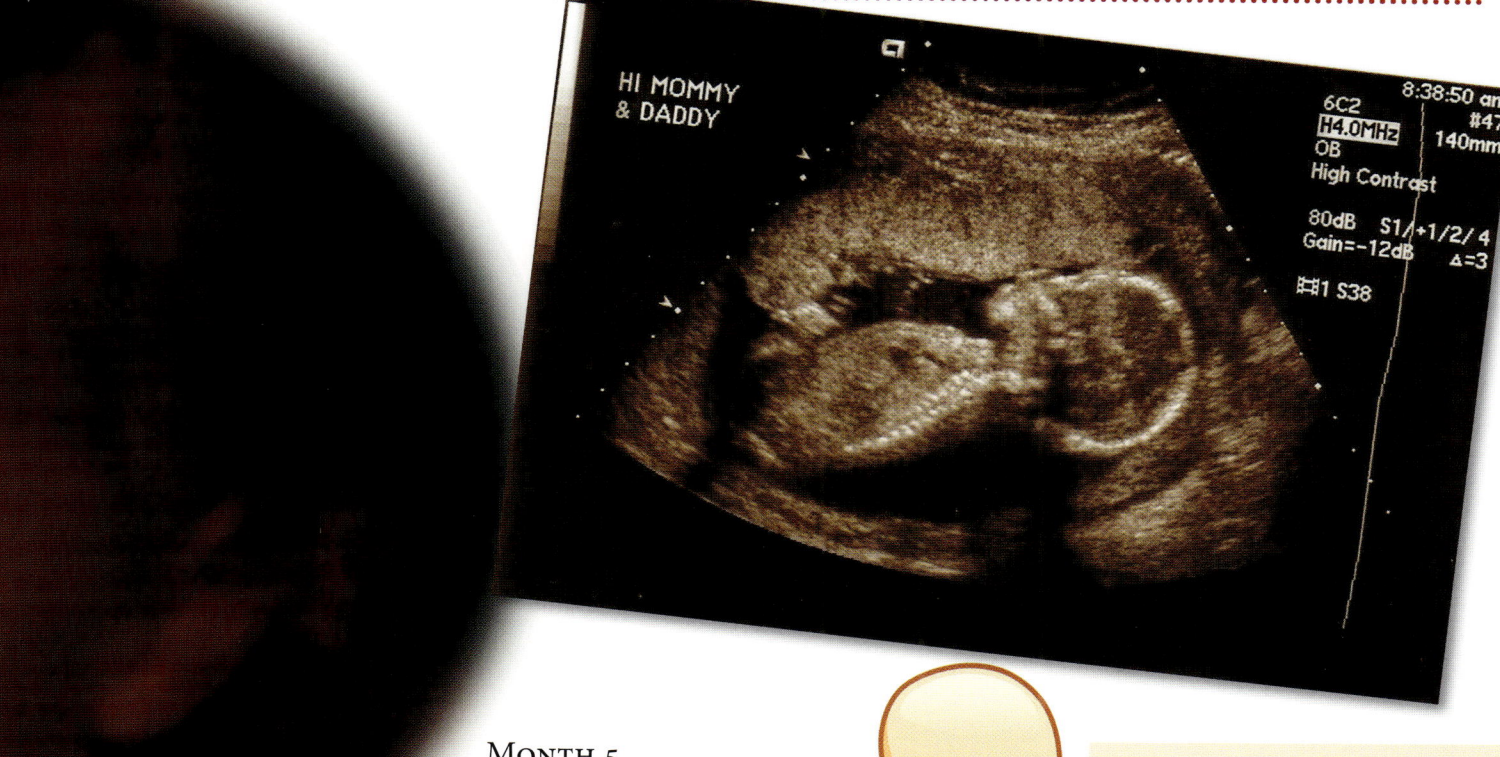

Month 5

Internal organs are maturing, eyebrows, lids, and lashes appear, and hair is finally growing on her head. Mom can now feel the baby's tiny movements like "fluttering butterflies" in her belly.

Did you know?

A baby's red blood cell factories are originally located in a sac outside the embryo in the womb. They eventually migrate to the liver and spleen before landing permanently in the bone marrow during the fifth month of gestation. When baby grows to adulthood, if the body is stressed by a shortage of red blood cells, the spleen and liver can actually reopen their factories to compensate.

Month 6

The baby may hiccup and occasionally open her eyes for a little while. *Vernix*, a protective cheese-like coating, "waterproofs" the baby's skin while she is submerged in the amniotic fluid.

Month 7

The baby gains taste buds, and she's loading up on healthy layers of fat. A baby born at this time could potentially survive outside the womb with special medical care.

Month 8

The brain develops extensively and the baby increases rapidly in size and weight. Lungs still need further development. Mom can feel the baby's strong kicks and see them from the outside of her body.

Month 9

Babies vary in size at the end of gestation, but length is usually around 19 to 20 inches and weight averages about seven to seven and a half pounds. The baby finishes development and settles lower into mom's pelvic region, getting ready for birth.

Did you know?

A baby thrives in a total water world for nine months—a world that is utterly impossible for any person to live in immediately after his very first breath. That feat is accomplished by the baby possessing—only in the womb—blood vessels with a different arrangement and structure than an adult's.

Interdependence: Proof of Purposeful Design

When studying the human body, we can all observe and recognize good design. We see it in the way blood clots to bind a wound or the way a mother's body makes the perfect food for her baby. Renowned evolutionist Richard Dawkins asserts that "biology is the study of complicated things that have the appearance of having been designed for a purpose" (*Blind Watchmaker*, 1). He recognizes that the human body "appears" to be designed but denies its reality. Is design just an illusion? One way to tell is by considering how the body's systems display the principle of interdependence.

INTERDEPENDENCE

The human body is full of interdependent systems—systems in which two or more parts mutually depend on each other for proper function. If more than one part is required for proper function, then how could one part evolve first and function before the rest? It's like the chicken-and-egg dilemma, only much more complex. Interdependence challenges the idea that systems could develop step by tiny step in a long, slow evolutionary process.

FEATURES OF A DESIGNED SYSTEM

1. Numerous parts working together for a purpose
2. A particular arrangement of the parts
3. Proper alignment
4. Precise timing
5. Exact dimensions and shape
6. Tight fit
7. Definite sequence for correct assembly

DNA, RNA, AND PROTEINS

Mechanisms inside cells display interdependence—particularly DNA, RNA, and proteins. DNA contains the blueprint of an organism and instructions on how, when, how often, in what amounts, and where to build new cells and proteins. Some RNA strands contain specific DNA information used to build proteins. Proteins are responsible for manufacturing more DNA within new cells. So DNA creates RNA which creates proteins which create DNA. None of them can exist on their own. It's one of the most fundamental interdependent systems of biological life.

DNA

RNA

Protein

A car is a man-made example of an interdependent system. How far could you drive if it had a motor but no wheels? What if it were only missing the steering wheel—would you make it home? These are silly questions because we all know the car could not possibly function properly without the cooperation of all of its main components. Yet evolution claims we came from organisms that continued to live and reproduce while waiting for all of their vital parts to evolve.

12 V
64 Ah

Heart and Placenta

Among the many interdependent systems that must come together to enable a pregnancy, some of the most important are the placenta and both the baby's and mother's hearts. The placenta secretes a hormone that regulates the cholesterol levels of the baby's heart, so the development of a human heart depends upon the existence and proper function of a placenta. However, the placenta requires blood circulation that the mother's heart supplies. Which came first? The heart could not develop in the womb without the placenta, but the placenta could not function without the prior existence of a heart. Both organs would have to be in place at the same time for any human to be born.

Abdominal aorta

Placenta

Umbilical vein

Umbilical arteries

Reproduction

Human reproduction requires functional, interdependent parts found in separate individuals. Explanations of how reproduction could have evolved fail when they appeal to the prior existence of a reproducing organism. Evolutionists will sometimes point to in vitro fertilization as an example, but that starts with the prior existence of a donor egg and sperm. How do organisms "arise" by increments until they can reproduce? Reproductive systems need all of their vital parts at once, which implies intentional design.

Heart, Lungs, and Digestion

In humans, the heart, lungs, and digestive systems would fail if any one of them were absent. The circulatory system depends on the digestive system to supply usable energy the heart muscle can burn as it pumps blood. Meanwhile, the digestive system depends on the circulatory system to distribute oxygen to its tissues. Without oxygen, digestive cells would die. The heart and digestive system both need the lungs to get rid of carbon dioxide waste. But the lungs need energy from processed food and the carrying capacity of the circulatory system in order to expel carbon dioxide. Such intricate interdependence shows these three systems must have existed together from the beginning.

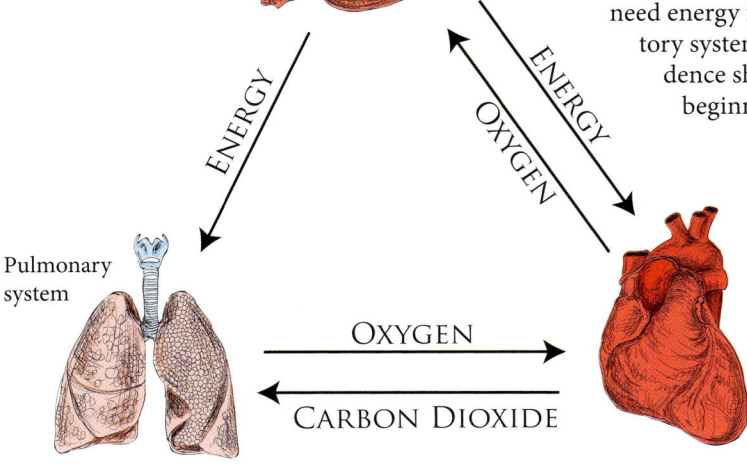

Digestive system

ENERGY

ENERGY

OXYGEN

Pulmonary system

OXYGEN

CARBON DIOXIDE

Circulatory system

If gradual evolution over eons of time cannot account for these interdependent systems, then what can? Purposeful design by an all-wise Creator provides the perfect explanation.

Why We Look Similar to Our Parents

Do you have your dad's brown eyes? Or your mom's red hair? How did that happen? Though you are similar to your parents, God created you to be a one-of-a-kind individual with a unique set of physical traits, abilities, and gifts. Many aspects of these attributes are based on heredity, the genetic inheritance you got from your parents. You received half your genes from your dad and half from your mom, so you have a mixture—a unique arrangement—of their genes. The specific combination you inherited makes you different from anyone else who has ever lived. Even so, you share some characteristics with your parents and other relatives because you share some of the same genes.

INHERITANCE PATTERNS

Almost all human traits arise from a complex mix of many hereditary patterns. For instance, why do some people in a family have dimples or freckles or a cleft chin and others don't? It depends on the particular genetic variations a person inherited. In a Mendelian inheritance pattern, the variation of a gene is called an *allele*. Some alleles are considered dominant and are more likely to be expressed—i.e., to show up in the final physical trait. Other alleles are recessive and less likely to be expressed. If a trait like dimples runs in your family and you received two dominant alleles for it from your parents, or even one dominant and one recessive, then you'll probably have dimples. But if you received two recessive alleles, you won't. However, most human physical traits follow more complicated patterns than the Mendelian one. For instance, genes are controlled by non-genetic information, one gene may influence many traits, and many genes may be needed to make one trait.

"And He has made from one blood every nation of men to dwell on all the face of the earth, and has determined their preappointed times and the boundaries of their dwellings." (Acts 17:26)

Did you know?

Nearly all genetic mutations are either neutral or bad. Instead of helping, they harm. Thousands of diseases and many cancers are linked to mutations. As they accumulate in a population over time, they may threaten a species with extinction. Studies on human mutation rates show we couldn't have been around for millions of years. We began only thousands of years ago, just like the Bible says.

Mutations

Cell division is the biological basis for how living organisms develop, function, and reproduce. When a cell divides, its DNA is copied into two new cells. Every division offers a chance for copying errors—called mutations—to happen. These mutations represent a loss or corruption of the previous genetic information. Mutations that occur in sperm or egg cells accumulate over time and are passed from parents to their children with each new generation. Some physical traits are linked to lost genetic information. For instance, people with blue eyes have a mutation that limits the production of melanin, the pigment found in dark eyes.

Could Adam and Eve Produce All These Different People?

When you look at all the differences among the peoples of the world—height, hair, eye color, skin color, and more—you might wonder how so much variety came from just two people. The answer is that God built variations into the genetic code and designed other mechanisms from the beginning. They didn't all arise by genetic mutation. Adam and Eve were created with the full set of genetic information and other elaborate mechanisms for all the humans who descended from them. Each of their children received unique combinations of those genes, as did their children after them, and so on. As people groups spread out after the Tower of Babel, they became geographically and socially isolated. This led to increased genetic isolation, which coupled with mutations to result in differing ethnic groups. One difference is skin color, which is determined by the amount of melanin in the skin. Adam and Eve could have started out with all the genes needed for every shade we see today. Each son or daughter inherited some combination of those genes, which they passed down to the following generation in a new combination. As the recombined genes became more removed from the original source, some people would have had a greater tendency for dark skin and others for light. Even with our differences, all human genomes are still 99.9% alike. As the Bible makes clear, we are all descended from Adam and Eve, and we are all of "one blood."

Growing Up and Growing Old

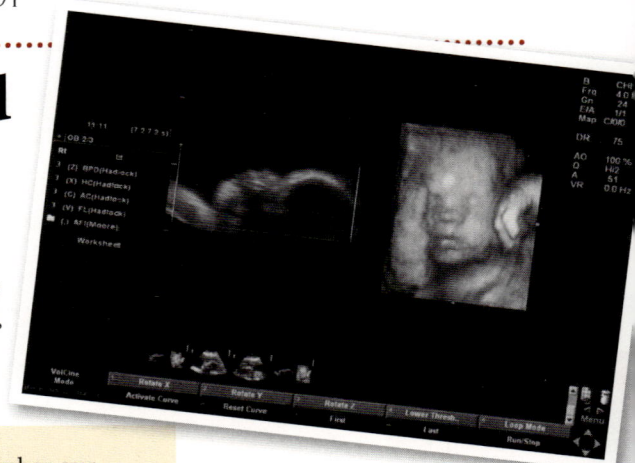

BABIES IN THE WOMB: NINE MONTHS

Life begins at the moment of conception. The developing human grows at a rapid rate over a period of nine months. Brain, organs, and body systems continually grow until the baby is full-term and prepared for life outside the womb. Inside the womb, a baby receives oxygen and nourishment from the mother via the placenta, a temporary organ. After baby is born, his lungs begin to provide oxygen to his body, and he is designed to receive nourishment from his mother's breast milk.

INFANTS: THE FIRST YEAR

The first year of life outside the womb is characterized by rapid development and weight gain. An infant learns the basic elements of language and builds the strength needed to sit up and eventually walk. He also begins to eat solid foods as several teeth grow in through the gums. An infant has separations between bones of the skull called *fontanels* (or "soft spots") to allow for skull expansion, accommodating rapid brain growth.

> "So teach us to number our days, that we may gain a heart of wisdom." (Psalm 90:12)

TODDLERS AND PRESCHOOLERS: AGES 1-4

Toddlers and preschoolers further develop their fine motor skills and athletic abilities, enabling them to run, hop, write, and draw. Language skills progress from speaking a few words by age one to constructing sentences by age two. Vocabulary grows and their understanding broadens. Personality traits emerge and are designed to be refined as they relate to other people and events, including whether a child will be introverted or extroverted. Their play also becomes more imaginative. They desire to have more control over themselves and their environment.

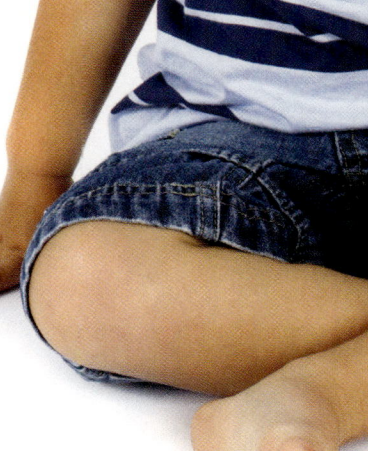

EARLY CHILDHOOD: AGES 5-9

In early childhood, brain and physical growth continues, with only occasional spurts in height. The ability to learn and memorize new information is at its peak during this stage, and the child has the capacity to speak fluently in more than one language. Up to this point, their concept of the world has centered around home. During childhood years, their perspective is broadened to other families, communities, and the world. Children learn limited independence and begin exploring their unique talents and interests. They develop habits and a belief system, some of which may carry on through the rest of their lives.

Late Childhood: Ages 10-12

During late childhood, puberty begins with an increase in hormone levels. Gender-specific physical changes appear that begin to turn bodies of boys and girls into adult men and women, preparing them to one day produce children of their own. Their command and understanding of language becomes more complex to include humor, metaphors, and sentences that can communicate double meanings.

Adolescence: Ages 13-19

During adolescence—also known as the teenage years— the rate of physical development increases, especially growth in height and weight. Adolescents seek a sense of identity that is influenced by spiritual understanding, appearance, skills, interests, parents, siblings, school, and friends. They tend to seek greater independence from their parents and invest more deeply in a close circle of friends. Adolescents also begin thinking about their future vocation and calling.

Early Adulthood: Ages 20-40

In early adulthood, the capacity for physical and mental performance is at its height, peaking at about age 30. For many, this is a physically ideal time for bearing children. As the body's metabolism slows, it tends to lose muscle mass and put on extra weight. Organ systems become less efficient and gray hair and wrinkles begin to appear during the latter part of this stage. But a healthy lifestyle can extend the time the body is able to function at its best. Wisdom and expertise develop as adults have lived long enough to learn some deeper lessons of life.

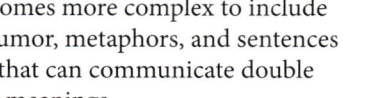

Middle Adulthood: Ages 41-65

In middle adulthood, senses such as hearing and vision begin to decline in performance. Women go through a time when their body transitions to a state where it can no longer bear children—a stage known as *menopause*. Reaction time and speed of thinking appear to slow a little, but knowledge of the world from years of experience provides its own advantage.

Late Adulthood: Age 65+

In late adulthood, organs and body systems are weaker and more susceptible to illness. Hair is mostly gray by this time, and skin wrinkles increase. The brain becomes smaller and appears to process more slowly. However, mental and physical exercise can extend function of the brain and body. In the United States, the average person lives to be in their mid-70s or early 80s.

Language: The Gift of Communication

There is no comparison between the complexity of human speech and the grunts, barks, and chatterings of animals. Although talking seems simple, it actually depends on so many perfectly placed characteristics—both material and immaterial—that language must be a gift from God.

THE HISTORY OF LANGUAGE

The Lord Jesus Christ created Adam and Eve and immediately communicated with them in language their created minds could understand. They and their descendants continued to use this created language, even speaking to God in prayer. But during the rebellion at Babel, the Bible says God confused the language of the people and scattered them over the face of the earth (Genesis 11:1-9). The people dispersed from Babel probably represented about 70 basic languages, judging from the 70 ancestral tribes listed in Genesis 10. Over time, these languages have multiplied into many others, and some have gone extinct.

> "For there is not a word on my tongue, but behold, O LORD, You know it altogether." (Psalm 139:4)

THE GIFT OF LANGUAGE

It was God who, as the eternal Word Himself, created the marvelous gift of human language. He designed the mouth, tongue, voice, and brain—every part you need to communicate with Him and with other people. God's gift of language reveals His Word and intentions for us and provides a way for us to respond in faith and praise to Him.

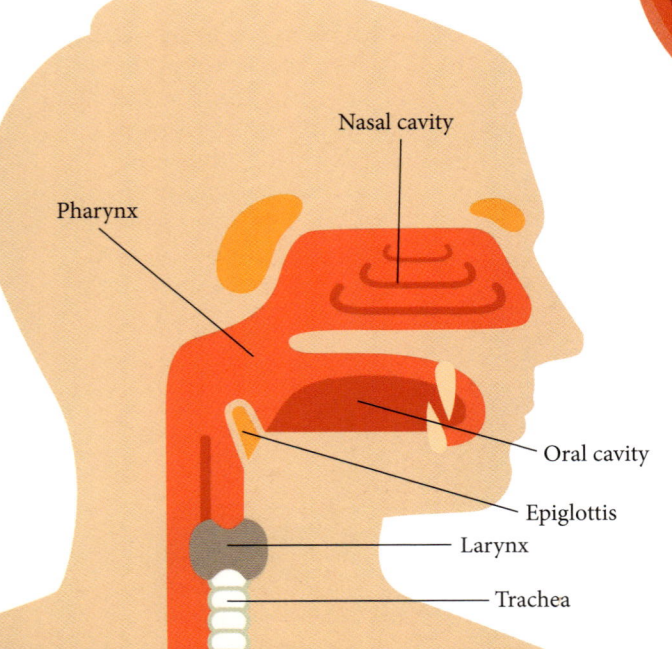

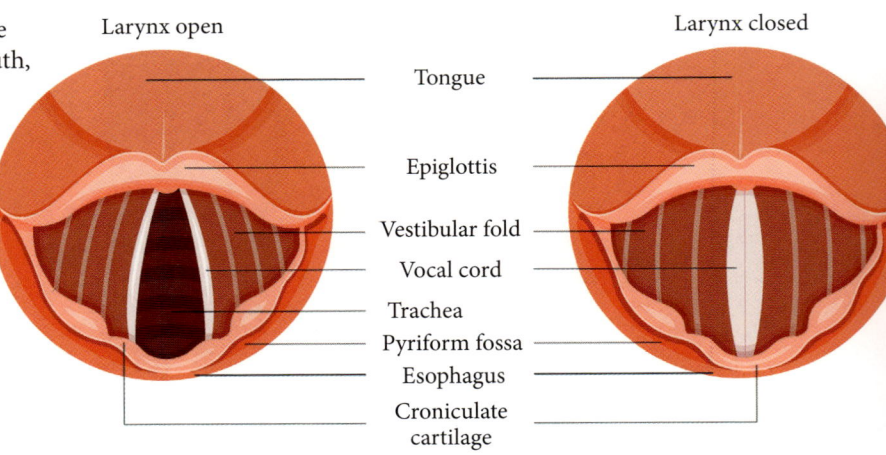

Larynx open

Larynx closed

Tongue

Epiglottis

Vestibular fold

Vocal cord

Trachea

Pyriform fossa

Esophagus

Croniculate cartilage

Nasal cavity

Pharynx

Oral cavity

Epiglottis

Larynx

Trachea

LANGUAGE NEEDS A VOICE

What produces your voice as you chat, yell, whisper, or sing? It all begins as your lungs send a powerful airstream into the trachea as you exhale. The larynx, also known as your voice box, sits at the top of your throat and receives the air from the lungs. Two folds of tissue in your larynx known as vocal cords or vocal folds vibrate together as the gust of air passes through. The vibrations can vary anywhere between 100 and 1,000 times per second. Specialized muscles control the length and tension of the vocal cords to determine the pitch of your voice. These vibrations then resonate into the throat, nose, and mouth, producing your unique sound. Specific sounds and words are formed by changing the shape of the throat, mouth, tongue, and lips.

Language Needs the Human Brain

Though a voice is vital for speaking, humans must also have a brain capable of using and understanding speech—and thankfully we do! Though recent research reveals that numerous parts of the brain coordinate together to enable speech, two primary regions are still believed to play a strong role. Broca's area of the brain regulates the ability to speak, while Wernicke's area aids understanding of speech.

Paul Broca

Did you know?

Adult humans cannot breathe and eat at the same time without choking. But the voice box is higher in a baby's throat, allowing her to breathe while she nurses. It descends at around nine months of age, where it will remain for life and provide a wider range of sounds needed for language.

Early Development of Language

Beginning around six months, babies' brains learn how to form the right sounds for language with coordination of their throat, lips, tongue, and diaphragm muscles. Understanding the structure of words, suffixes, and tones comes next. Vocabulary soon follows. Toddlers at the peak of their language learning can add 10 to 15 new words per day. After several months of learning the basics, their brain transitions out of this intensive language-learning mode. No other creature has a stage of language development like this. Researchers continue to investigate the causes of this miraculously designed period.

Hyoid Bone

One physical feature that enables human speech is the hyoid bone that anchors our voice box. Apes do not have this bone. Often, evolutionists claim a certain fossil represents a human ancestor, but without a hyoid bone, it could be just an extinct kind of ape.

Hyoid bone

"I will sing of the mercies of the LORD forever; with my mouth will I make known Your faithfulness to all generations." (Psalm 89:1)

Genetic Entropy and Glorified Bodies

Ticking within every species is a "genetic clock" that marks the length of time a species has existed on Earth. Mutations steadily accumulate within a species. These clocks indicate that life is thousands of years old, not millions, and they also show that our genes are breaking down as time goes by. Adam and Eve were originally created with perfect, error-free genes—no mutations were present in their bodies. When sin entered the world, the whole creation became cursed—and that includes our DNA. The human genome has been on a downhill slide ever since the Fall, accumulating mutations with each successive generation.

This is not good news for the human race. But take heart! There is hope.

"Even though our outward man is perishing, yet the inward man is being renewed day by day." (2 Corinthians 4:16)

GENETIC ENTROPY

Researchers modeled the accumulation of mutations in the human genome over time using computer simulations. They found the buildup of mutations can only reach a certain level before the genome completely deteriorates and humans go extinct. This degradation, called *genetic entropy*, fits perfectly with a recent creation of six or so thousand years ago.

The genetic entropy trend also fits the pattern of human lifespan decline after the Flood as recorded in the Bible.

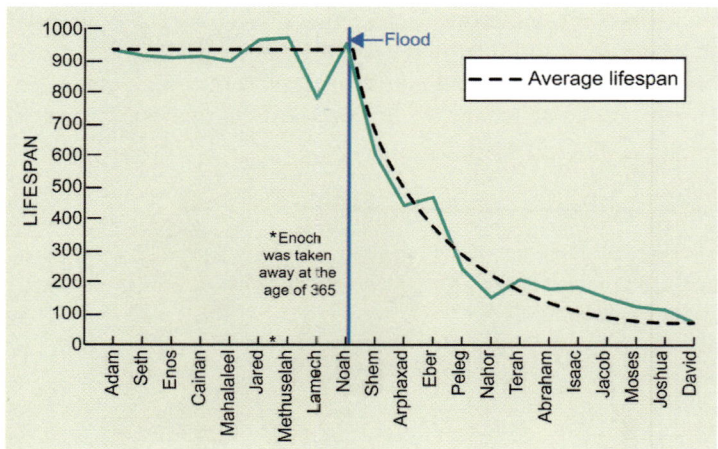

A mutation is like a typing error in our DNA. It becomes a permanent change in genetic sequence when our body's repair systems do not readily correct it. Each generation adds to the total number of "typos" in the genome, so the total number of mutations in mankind's genome builds up over time.

We are not gradually evolving better genomes as Darwinism teaches. Instead, mutations occur regularly and are far more likely to be harmful than helpful.

BIBLICAL TIMESCALE

Widely accepted evolutionary conjectures assert that mankind evolved from ape-like ancestors about three to six million years ago. Humans could have experienced 120,000 generations within that amount of time. In contrast, Scripture tells us about 200 generations have passed since creation.

ENTROPY OF THE BRAIN

Even some secular researchers are beginning to recognize that, contrary to Darwinian expectations, genetic entropy affects our brains. Human intelligence relies on neurons, and these cells work best if their genes stay in top shape. But like all DNA, neuron genes decay. The resulting errors pile up with each successive generation and begin to diminish intelligence.

HAS OUR INTELLECT PEAKED?

Our brains use 2,000 to 5,000 so-called *Intellectual Deficiency* (ID) genes. Geneticists routinely identify specific mutations in ID genes as causing various types of intellectual deficiencies. Each generation accrues about 100 new mutations to the gene-coding DNA regions of the human genome. At this rate, each person should have accumulated a half dozen mutations in ID genes in just the last 3,000 years. Evolution's defenders face a challenge when trying to explain why humanity's ancient ancestors never lost their intellect even after a supposed few million years of mutations to ID genes.

MITOCHONDRIAL DNA

Genetic data from DNA in the cell's mitochondria also demonstrate genetic entropy in a recent creation. Mitochondrial DNA is inherited from our mothers, and geneticists have measured its mutation rate. At this rate, after 2.4 million years, all the mitochondrial DNA should have been changed by mutation several times over. But 10,000 years would produce only about 38 mutations. Mitochondrial DNA sequences between individuals show about a dozen differences, demonstrating that genetics strikingly confirms the world is thousands, not millions, of years old.

> "For our citizenship is in heaven, from which we also eagerly wait for the Savior, the Lord Jesus Christ, who will transform our lowly body that it may be conformed to His glorious body." (Philippians 3:20-21)

THE BRAIN AND THE BIBLE

Genetic entropy confirms three Bible teachings. First, Adam and Eve's brains were originally optimal and without genetic defect. Second, we had our best brains about 6,000 years ago. Third, humanity has suffered genetic degradation since the Genesis 3 Curse.

We marvel at God's engineering genius as we look at how wonderfully we are made. Now imagine the glorious creations Adam and Eve must have been before the Fall. No hope exists in nature or science to escape the results of genetic entropy, but in the Bible God promises a better future with new and far more glorious bodies. For those who turn from sin—the source of our decline—and put their trust in the Savior and Creator, the Lord Jesus Christ, entropy is temporary and glory is eternal.

Confronting Cancer

Cancer is one of the most serious health problems a body can face. While cures for cancers remain to be found, researchers are learning more about it every day. And with over a hundred known types of cancer in existence, there is plenty of research to be done.

What Is Cancer?

Cancer begins with a genetic malfunction in a single cell. Research has revealed that when controls over cellular growth break down in a cell, the cell starts replicating without restrictions. Then the "daughter" cancer cells replicate wildly. In addition to this abnormal, out-of-control cell growth, the body begins to accumulate dead cells instead of discarding them. Tumors form as a result. If other malfunctions block the body's defensive response to tumors, the cancer cells will spread unhindered. Since cancer is a genetic disease that occurs in normal cells, it can show up in anyone without warning. But tobacco use, an unhealthy diet, physical inactivity, old age, and the overuse of alcohol can increase your chances of developing cancer.

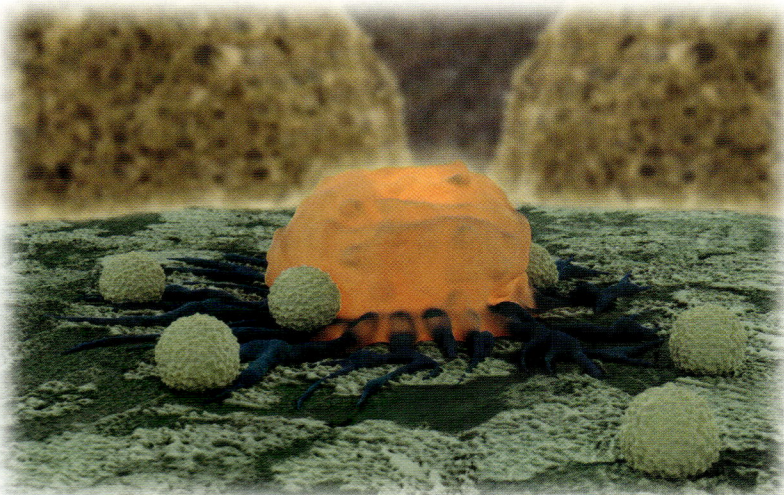

When Cancer Spreads

When cancer spreads to other parts of the body, it is referred to as metastatic cancer. Cancer cells break away from their growth sites and travel through the lymph system or bloodstream to lymph nodes or organs, where they can form new tumors. Everywhere the cancer cells spread, they outgrow normal cells. When too many normal cells are destroyed, that body function will cease. Tumors may erode through tissue barriers and blood vessels. Very large tumors may severely compress organs, leading to a loss of function.

Carcinoma

Carcinoma is the most common type of cancer. It forms through DNA damage to epithelial cells like skin. Squamous cell carcinoma is a cancer that forms from a kind of epithelial cell called a *squamous cell.* These cells lie just beneath the outer surface of the skin and envelop organs such as the stomach, intestines, kidneys, and bladder.

Leukemia

Leukemia does not develop tumors but rather increases the production of white blood cells so much that they clog the bone marrow and suppress the production of normal red blood cells. The low level of red blood cells makes it hard for the body to distribute nutrients and oxygen. Though the number of white cells increases, they do not function normally. So a person is at a higher risk of infection.

Lymphoma

Lymphoma develops in lymphocytes, white blood cells that regulate microbes as part of your immune system. Several types of lymphoma exist, but all can fall into one of two categories: Hodgkin's or non-Hodgkin's. Which type a person has is determined by what type of lymphocyte the cancer involves. These white cells also don't function normally, so a person is more susceptible to infection. Large numbers of these cells form tumors in lymph nodes.

Sarcoma

Sarcoma cancer forms in bone and soft tissues of the body, such as muscle, tendons, fat, blood vessels, lymph vessels, nerves, and joint tissue. Most of the time sarcoma cancer tumors develop in the arms or legs.

Melanoma

Melanoma develops in the melanocytes, skin cells that produce melanin and give your skin its color. It is the most serious form of skin cancer because it can spread so quickly. Risk factors include level of sun exposure, number of moles on the skin, skin type, and genetics.

DIAGNOSING CANCER

Most cancers are diagnosed through a routine checkup for a mild symptom like abdominal pain. However, a tumor is usually a non-painful, very hard mass. Other symptoms include difficulty swallowing, blood in urine or feces, a constant fever or cough that lasts longer than three to four weeks, or strange skin spots. If cancer is suspected, then the patient's medical history is examined, followed by laboratory studies of blood, urine, or stool. If a tumor is suspected, then imaging tests such as MRI, X-ray, or ultrasound are used. If these examinations are positive, then the final confirmation is made through a biopsy or surgery.

CANCER TREATMENTS

Treatments will vary depending on the kind of cancer and how much it has spread. Chemotherapy includes the use of powerful chemicals that destroy very rapidly dividing cells, which include cancer cells. Unfortunately, the chemicals also attack healthy cells, leading to various side effects. Radiation therapy uses high-energy radiation to damage the DNA of rapidly dividing cells like cancer cells. Radiation may be delivered from an outside machine or radioactive material placed inside the body. Though radiation can be an effective tool, it too may damage surrounding healthy cells. Doctors must monitor cancer patients closely during all of these treatments to ensure their bodies do not become too weakened in the process.

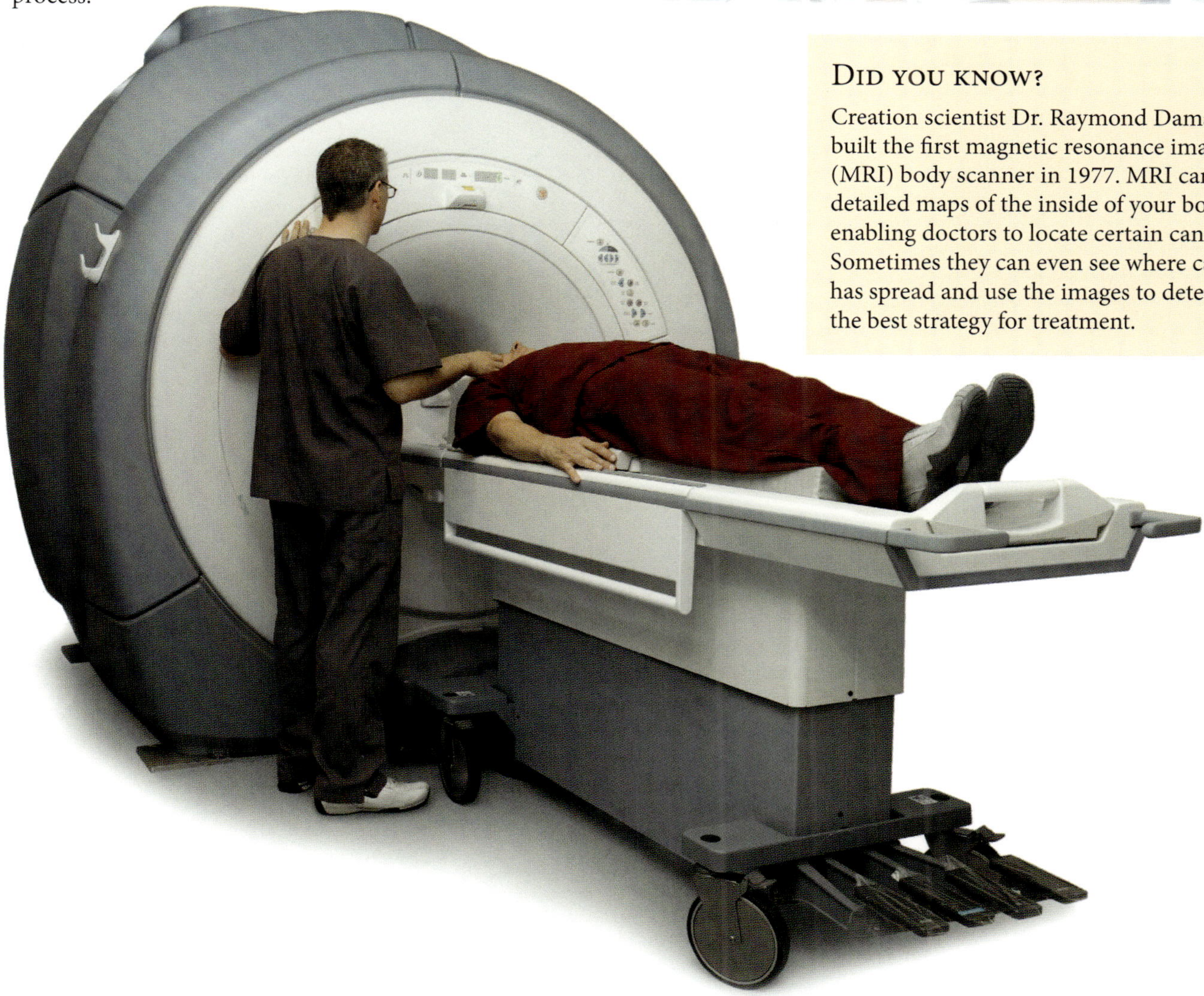

DID YOU KNOW?

Immunotherapy involves stimulation of the immune system in a variety of ways to boost the body's ability to fight cancer with its natural defenses.

DID YOU KNOW?

Creation scientist Dr. Raymond Damadian built the first magnetic resonance imaging (MRI) body scanner in 1977. MRI can create detailed maps of the inside of your body, enabling doctors to locate certain cancers. Sometimes they can even see where cancer has spread and use the images to determine the best strategy for treatment.

What Makes Us Sick?

God designed the human body to be strong, but people worldwide are susceptible to sickness—from mild cases of the sniffles to serious illnesses. Although God created the world as a very good place, when sin entered the picture, so did death, disease, and physical corruption. Some health threats come from within, like cancer, but others come from outside the body. What are some of the things that can make us sick?

BACTERIA

Our world is teeming with bacteria, tiny organisms we can only see through instruments like microscopes. Bacteria can be found in almost every environment. Most are helpful; in fact, they are essential for life. Some strains of bacteria, however, now cause sickness. These are called *pathogens* and are a major cause of disease and death. Antibiotics are used to either kill or slow the growth of pathogenic bacteria. Until antibiotics were discovered, even the smallest infection could prove deadly.

DID YOU KNOW?

Certain bacteria can develop resistance to the antibiotics used to fight them. Some people say this is evidence of upward evolution, but actually these "superbugs" either acquired a gene from other bacteria or experienced a genetic mutation that gave them a resistance to the drugs at a cost of the reduced efficiency of some cellular process. Either way, their information content goes downhill, which is the opposite of what evolution requires.

DID YOU KNOW?

One of the most well-known bacteria is *Escherichia coli* (*E. coli*), which can cause serious food poisoning. *E. coli* was used in a 20-year experiment to see what kinds of genetic changes would occur. After 50,000 generations, the bacteria wound up normal, mutant, or dead—none evolved.

PARASITES

People are also susceptible to parasites—worms and other creatures that can invade and live off the human body. Some people are infected when they wade or swim in water where the parasites live. Other invaders can come through food or contact with insects and animals that carry the parasites. Once a parasite sets up shop, it reproduces within the body and can cause infections and other maladies. Not all parasites are harmful, though. Some even help. A certain whipworm has been used to treat colitis, an inflammation of the colon. Perhaps some now-parasitic worms were originally designed to maintain intestinal health.

Model of a Tapeworm

VIRUSES

Viruses cause some of the most devastating diseases humans experience, such as smallpox, polio, and Ebola. These small infectious agents can only replicate inside the cells of living creatures. They are like tiny machines that inject DNA or RNA into a living cell. The cell duplicates copies of the virus in very large numbers that can cause sickness. Some scientists believe that viruses aided evolution by introducing new genetic material that cells could adapt into new features. However, not only can they not explain how this process happened, research shows that what they thought were viral genetic "junk" leftovers are actually functional parts of the genome.

Electron micrograph of the Ebola virus

DID YOU KNOW?

Engineers who design water treatment plants and sewage processing facilities that keep dangerous bacteria, viruses, and parasites away from people save more lives in a year than medical doctors.

POISONOUS CREATURES

Some land and sea creatures can also make us sick. These include fish such as the lionfish, many kinds of venomous snakes, certain spiders and scorpions, some jellyfish and other sea creatures, ticks and fleas that carry infections, and stinging insects that can cause allergic reactions or even kill if someone is stung too many times. If God originally created the world without sickness and death, where did the dangerous features of these creatures come from? Scientific studies suggest that their poisonous features originally served a different purpose. They became corrupted after the Curse so that what was once benign or helpful is now harmful in certain circumstances. Meanwhile, many creatures' toxins are used as medicines when applied in appropriate doses and places in the body.

DID YOU KNOW?

One of the most common human parasites is the intestinal tapeworm, which is often acquired from undercooked meat or food that wasn't prepared in clean conditions. So always be careful what you put in your mouth!

CPR: A Real Lifesaver

Picture this: You're at a picnic in a friend's backyard. You've just filled your cup with pink lemonade when the elderly man behind you falls to the ground clutching his chest and then goes limp. You fear he had a heart attack. Someone calls 911, but the paramedics will not arrive for several minutes. This situation is why the American Heart Association encourages people to learn CPR.

Cardiopulmonary resuscitation (CPR) is relatively simple, yet it could mean the difference between life and death. The American Heart Association guidelines are the basis for courses in CPR and are regularly updated as research findings indicate the current best methods. In general, CPR is a combination of techniques used to help the heart circulate blood and provide oxygen to the brain until further medical interventions can be applied. It can be very beneficial to take a CPR certification class to get hands-on experience and learn how to do it properly.

THE BASICS

1. If unresponsive, call 911.
2. Perform chest compressions.
3. Carefully follow the instructions if you use an automated external defibrillator (AED).

FIRST: CHECK RESPONSIVENESS

The first thing to know is whether the victim is conscious or not. Tap or shake him gently to try and get his attention. Ask loudly, "Are you OK?" If you don't get a response, then tell someone to call 911 as you begin CPR. If you are alone, call 911 before beginning CPR. If you are alone and the person is unconscious from possible drowning, perform CPR for one minute before calling 911.

SECOND: PERFORM CHEST COMPRESSIONS

In most cases, performing resuscitative breaths is no longer considered absolutely necessary. If you're untrained, don't worry about it. It's more important to keep the person's blood flowing since it's still filled with oxygen from his last breath.

Move the person so that his back lies on a firm surface. Kneel next to his shoulders. Place the heel of one hand in the center of his chest, between the nipples. Put your other hand on top of the first, interlocking the fingers. To perform a compression, lean over the person's chest, straighten your arms, and push down at least two inches. You're literally pumping the person's heart, so make sure it compresses enough to pump blood. Compress at a rate of 100 times per minute—almost twice a second. Allow the chest to rise completely between compressions. Perform until medical help arrives.

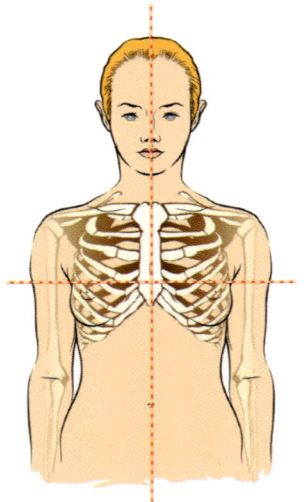
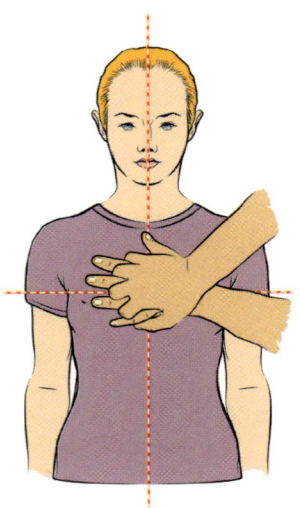
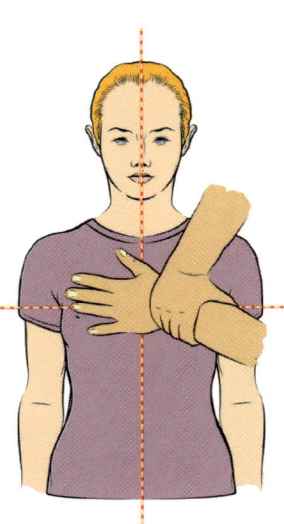
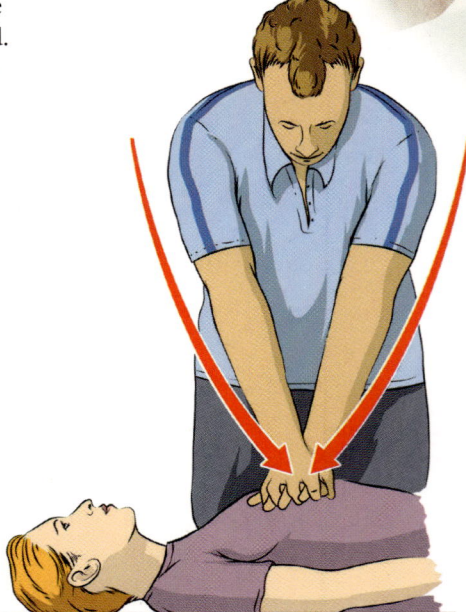

CPR ON A BABY

Performing CPR on a baby (0-12 months) is similar to the process on an adult. First, check for response by gently stroking or tapping his feet. If the baby does not respond, then call 911 and start chest compressions. Make sure the baby is on a firm surface, like a table or floor. Place two or three fingers in the center of the baby's chest between the nipples (do not use the heel of your hand) and compress about 1.5 inches. Do this at a rate of 100 compressions per minute. After 30 compressions, tilt the head back to clear the airway and give two resuscitative breaths, each lasting one second. Make sure the baby's chest rises. Continue the cycle of 30 compressions and two breaths until medical help arrives.

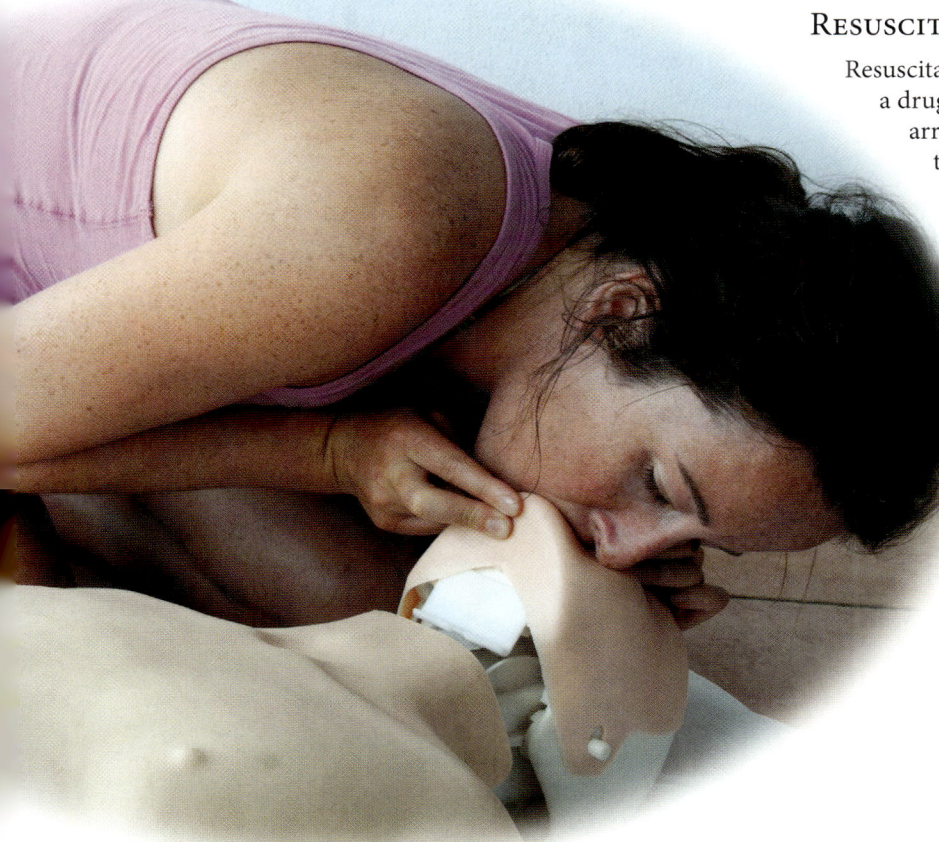

RESUSCITATIVE BREATHS

Resuscitative breaths should be used if the person suffered a drug overdose, near-drowning, or unwitnessed cardiac arrest. In these cases, perform 30 compressions, then two resuscitative breaths, and repeat the cycle until medical help arrives.

To perform a resuscitative breath, tilt the person's head back and lift the chin. This opens the airway. Lean close and listen for normal breathing. Gasping is not considered normal. If the person is not breathing normally, then continue with resuscitative breaths. Pinch the nose closed, completely cover the person's mouth with your own and make it as airtight as possible, then exhale for one second. Make sure the chest rises. If it doesn't, check the airway to see if it's open. You can use the acronym CAB to remember the cycle: compressions, airway, breathing.

AED

When available, an automated external defibrillator (AED) should be used to shock a stopped heart back into circulation. When you turn on the AED, it should provide instructions. Pediatric pads should be used for children ages one to eight, when possible. Experts recommend that an AED should not be used on an infant less than one year of age.

CAB

1. Compressions
2. Airway
3. Breathing

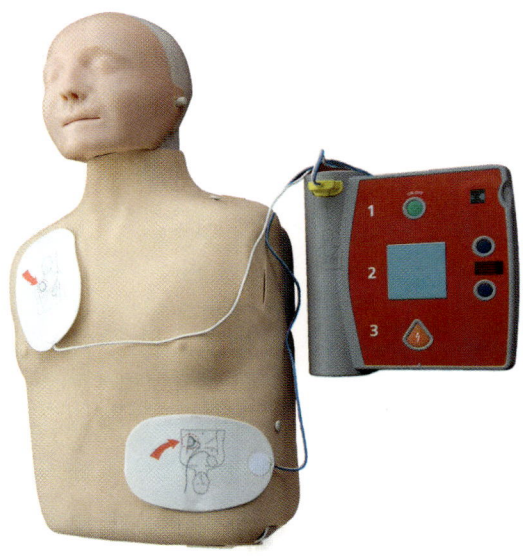

Life-Saving Organ Transplants

Transplanting organs is a remarkable medical procedure. A team of surgeons removes a healthy organ from one person and painstakingly inserts it into the body of another to replace a failing organ. They carefully connect the new organ's veins, nerves, and other tissue to the recipient's. For people who have major organ failure, a transplant is often the best option to save their life.

WHAT CAN BE DONATED?

Transplantable organs include the heart, kidneys, liver, lungs, pancreas, and intestine. Tissues such as corneas, bone, heart valves, skin, blood-forming stem cells, and blood vessels can also be transplanted. The organs and tissues from a deceased donor can save as many as eight people's lives! But a healthy person can also be a "living donor" by donating portions of an organ or by donating tissues the body can replenish. For example, a person who has two kidneys can donate one and still live using the one that remains. Also, the body of a person who donates blood, stem cells, or skin will simply make more to replace what has been given away.

Human kidneys

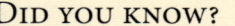
Heart being prepared for transplant

DID YOU KNOW?

The most commonly transplanted organ is the kidney. It is also the most successfully transplanted major organ, with a five-year survival rate of nearly 90%.

Many who need organs are placed on a national waiting list because of the limited supply of donor organs available. Some people who need new organs literally die waiting. The kidney is the organ in highest demand, and the most common tissue transplant is the cornea. Survival rates for organ transplants are high today, but this success is a recent development.

EARLY ATTEMPTS

Doctors in India grafted skin as early as 800 B.C. to treat burns, but true organ and tissue transplants have only been around about a century. In 1905, Austrian ophthalmologist Eduard Zirm performed the first corneal transplant, restoring the sight of a man blinded in an accident.

Eduard Zirm

ORGAN REJECTION

In 1933, Dr. Yuri Voronoy transplanted the first human kidney using an organ from a deceased donor. The transplant was initially successful, but the patient died from organ rejection. This case illustrates one of the major challenges for organ transplantation: without anti-rejection drugs, the recipient's body will almost always reject the donor organ because it has different "self" markers and the recipient's systems see the new organ as "non-self."

TRANSPLANT SUCCESS

Organ rejection was the norm in the early years of transplant medicine. The first successful kidney transplant between living patients occurred in 1954. The recipient and donor were identical twin brothers, and there was no immune rejection of the organ. The surgeon who performed the transplant, Dr. Joseph Murray, won the Nobel Prize.

Yuri Voronoy

THE DOMINO CHAIN DONATION METHOD

In 2005, Baltimore's Johns Hopkins Hospital developed the "domino chain" method of matching living donors and recipients. Willing donors who are genetically incompatible with their chosen recipients are matched with strangers; in return, their loved ones receive organs from other donors in the pool. A recent organ transplant chain set a world record with 60 patients and 30 kidneys!

ANTI-REJECTION DRUGS

In 1960, Peter Medawar won the Nobel Prize for his work in the development of anti-rejection drugs. Dr. Christiaan Barnard performed the first successful human-to-human heart transplant in 1967. The patient lived only 18 days—not because the transplanted heart failed or was rejected, but from pneumonia. The patient acquired this illness because his immune system was compromised by the anti-rejection drugs. In order to fight rejection of the donated organ, the patient must take anti-rejection drugs that compromise the immune system and increase the risk of patient infection. In addition to this side effect, the anti-rejection drugs are expensive and usually must be taken for the rest of the patient's life.

Peter Medawar

As transplants became more common, the U.S. Congress passed the National Organ Transplant Act in 1984, establishing a centralized registry for organ matching. You can become an organ donor by enrolling in your state's donor registry—often this is a choice you can make when you renew your driver's license.

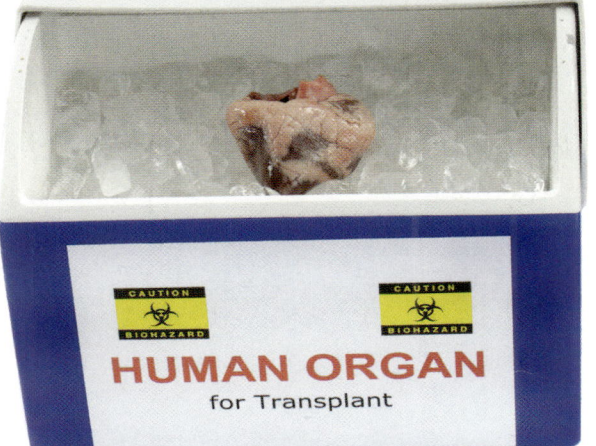

DID YOU KNOW?

Some people pre-register to donate their body and organs for scientific research, organ donation, or medical training after their death.

Poor Design and Purposeless Parts?

Evolutionary philosophy claims that the creatures we see in the world today resulted from millions of years of natural selection's trial and error. Because of this belief, many evolutionists contend that the human body is littered with "useless" (vestigial) or "poorly designed" organs left over from evolutionary tinkering. They also claim the presence of these organs refutes the existence of a designer. What omniscient, omnipotent creator would invest effort in parts without purpose? However, if scientists found evidence of poor design, it would not actually prove that humans evolved. It would only show that good design was corrupted. Even bad design is still design. But is there such thing as "bad design" within the human body? Or purposeless parts? We can find some answers by studying several of the organs that were once believed to be just that.

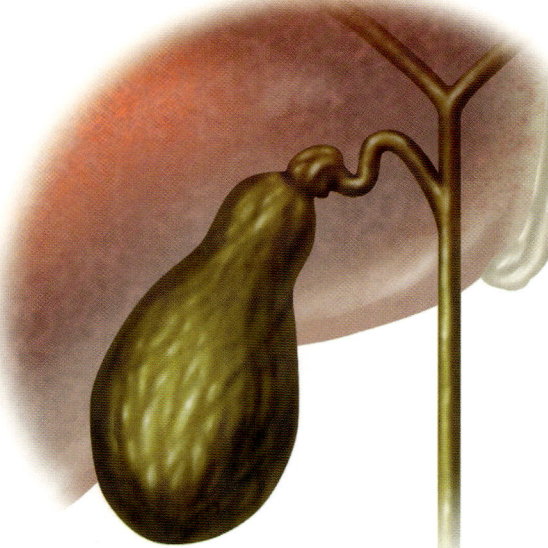

GALLBLADDER

A person can function without a gallbladder, but that doesn't mean it has no purpose. The gallbladder acts as a bridge between the liver and the small intestine. It receives and stores bile from the liver. When needed, the gallbladder expels concentrated bile into the small intestine for proper digestion. Without the gallbladder, bile simply trickles continually into the small intestine. While digestion is still possible without the gallbladder, it's not as efficient.

WISDOM TEETH

Wisdom teeth were once thought to be primitive remnants of a time when mankind had wider, thicker jaws. Supposedly, man has evolved to the point where his jaws have shrunk and no longer have room for all 32 teeth. So we surgically take some out. However, probably the biggest reason our wisdom teeth have trouble coming in is because of diet, not evolution. Modern processed foods do not promote the robust jaw development that consumption of crunchy raw foods provides.

APPENDIX

The appendix is around four inches long and sticks out from the large intestine. It can become inflamed, blocked, or infected, and so is sometimes removed. However, when it's healthy, the appendix has multiple uses in both the immune and digestive systems. God strategically located it at the junction of the small and large intestines.

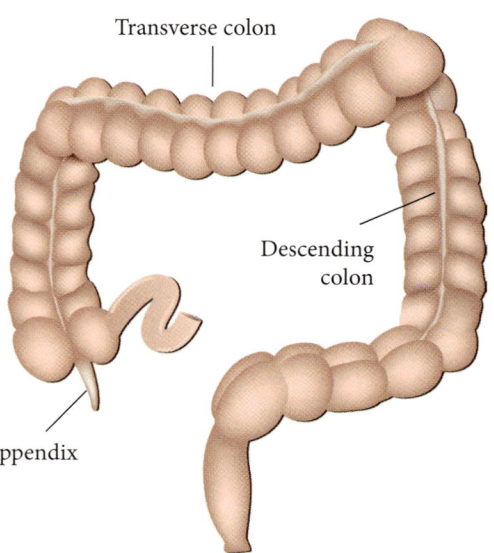

Transverse colon

Descending colon

Appendix

TONSILS

Like the appendix, tonsils can become infected and are surgically removed when necessary. At first, removing them didn't seem to have any health effects, so the medical community deemed them useless. However, tonsils are now known to sample orally ingested microbes. You can live without tonsils, but it's better to keep them around to properly regulate the microbes in your body.

Tonsils

TAIL BONE

The misleading name "tail bone" was given to the coccyx by those who considered it to be a residual feature from man's descent from tail-bearing animals.

However, it is now seen to perform important functions in muscle placement for the pelvic floor. It acts as an anchor for the anal sphincter muscles—so you wouldn't be able to relieve your bowels in the highly controlled manner you want without your tail bone.

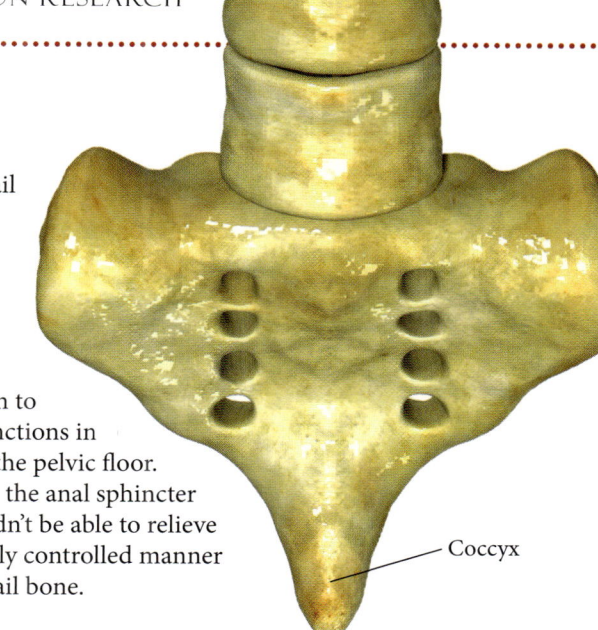

Coccyx

RECURRENT LARYNGEAL NERVE

The recurrent laryngeal nerve (RLN) connects the brain to the larynx and allows us to speak. However, evolutionists constantly point to this nerve as a hallmark of "bad" design. Instead of running a few inches from the brain to the larynx, it loops down through the neck, around the aorta, and back up along the esophagus to finally reach the larynx. This route actually serves three purposes.

1) While developing in the womb, every human starts out in the form of a tiny group of cells called a *blastocyst*. The RLN must remain functional through the elongation of the blastocyst and the addition of other body parts, tissues, organs, nerves, and vessels. During this period of development, the RLN actually functions like a tether to keep the heart in place instead of allowing it to float freely while the rest of the body is forming.

2) When a baby is born, the RLN performs more functions. The *left* one serves as a backup to the *right* one, and they both work to fine-tune vocal functions—so having two is not redundant, it's necessary.

3) Also, the left RLN does not loop around the aorta because it is "poorly designed." Rather, the RLN must take that route because it supplies the heart with nerves as well as the esophagus and trachea as it ascends to the larynx.

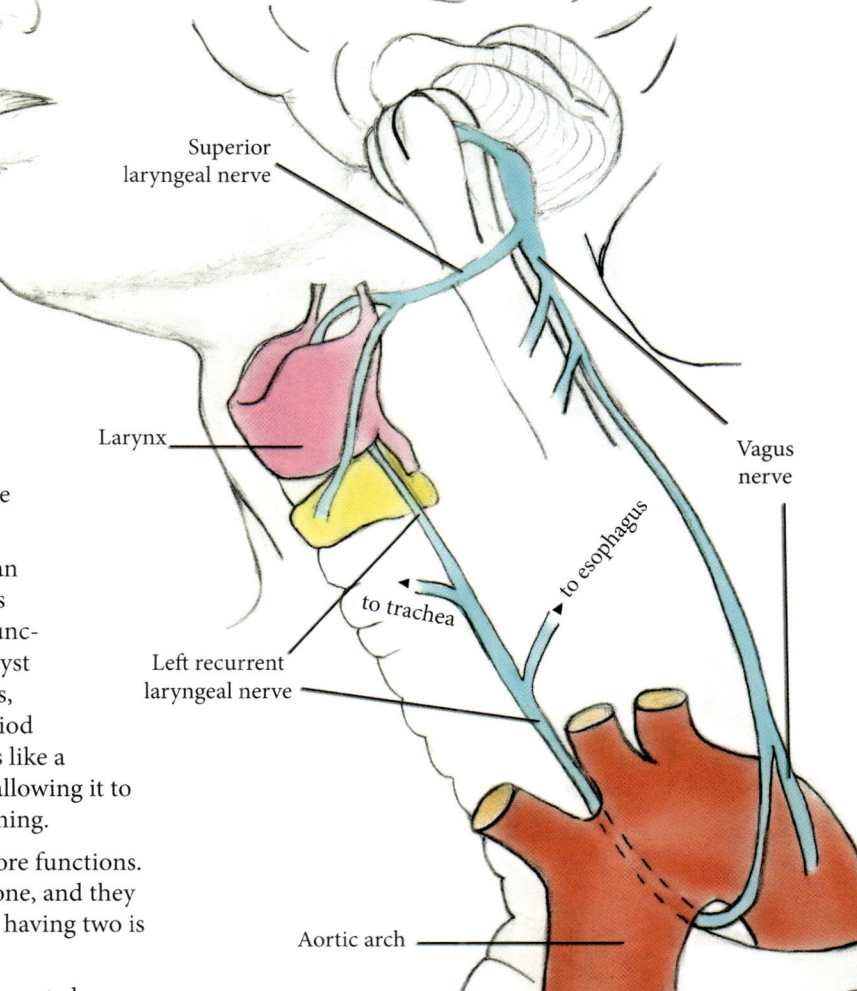

Superior laryngeal nerve

Larynx

Vagus nerve

to esophagus

to trachea

Left recurrent laryngeal nerve

Aortic arch

Testing the Limits

With training, the human body is designed to reach incredible levels of skill and athleticism. For thousands of years, people have displayed God's grand design when they challenge their body's strength, endurance, grace, and precision in a variety of sports and exceedingly difficult feats.

Swimming

Swimming may seem simple: just jump in the water and paddle. Yet this sport has come to include many different swimming styles and events. You can backstroke, breaststroke, do the butterfly, or even do competitive diving. Swimming is a low-impact sport, meaning that it's easier on the joints than running, but it also builds muscle mass and strengthens your heart. It can burn anywhere from 500-600 calories per hour.

DID YOU KNOW?

In 1875, Matthew Web was the first to swim the English Channel. He made it in 21 hours and 45 minutes, despite challenging tides and a jellyfish sting.

DID YOU KNOW?

Riders in the three-week-long bicycle race Tour de France lose weight because they cannot consume enough calories during the short time they're not riding or sleeping.

Dance

Every culture has developed its own style of dance. There's flamenco, swing, country, ballroom, ballet, Indian classical, Irish stepdance, and many others from around the world. Dancing can improve the condition of your heart and lungs, as well as increase muscular strength and endurance. It can help you age better and improve mental functions like memory.

Gymnastics

Gymnastics can be traced all the way back to ancient Greece, where it was originally intended for military training. However, like swimming, it has developed into many different events. Men's events include floor exercises, pommel horse, still rings, vault, parallel bars, and the high bar. Women's events include the vault, uneven bars, balance beam, and floor exercises.

ROCK CLIMBING

The goal of rock climbing is to get from point A to point B without falling. It sounds easy but is often quite difficult. Climbing requires agility, flexibility, and great strength in the arms and legs. Beginners use a technique called top-roping, which means they have a rope attached to them that runs to the top of the climbing area. This is what climbing walls have in parks and recreation centers. More advanced climbers tackle natural ledges and boulders, anchoring their safety rope as they ascend.

TRACK AND FIELD

The ancient Greek Olympics originally just featured a race from one end of the stadium to the other. Now it involves races of various lengths, relay races, hurdles, long jump, triple jump, pole vault, discus throw, javelin throw, hammer throw, and more. Records for these events used to only be recognized if they were performed outside, but now they can be held indoors as well.

THE MARATHON

The marathon originated from a legend of ancient Greece. As the story goes, the Greeks had just defeated the Persians in the Battle of Marathon. To prevent the Persians from sailing to Athens and falsely claiming victory so that the city would surrender, a Greek messenger ran almost 25 miles to Athens without stopping. He arrived and yelled, "Victory!" then collapsed and died. The human body can run this distance without dying but only after training. Marathons since that time have commemorated this legendary feat, but the length was extended to 26.2 miles in 1908.

WORLD RECORDS

100-Meter Freestyle Swimming: César Cielo, 46.91 seconds, 2009

Pole Vault: Renaud Lavillenie, 6.16 meters, 2014

Javelin Throw: Jan Zelezny, 98.48 meters, 1996

100-Meter Dash: Usain Bolt, 9.58 seconds, 2009

Discus Throw: Gabriele Reinsch, 76.8 meters, 1988

Marathon: Dennis Kipruto Kimetto, 2.02.57 hours, 2014

One-Mile Run: Hicham El Guerrouj, 3.43.13 minutes, 1999

OLYMPICS

Olympic athletes know all about testing the human body's limits. They must train rigorously to perform in challenging sporting events among the best international competitors. The Olympics were first held in ancient Greece. Competitions took place in the city of Olympia from the eighth century B.C. to the fourth century A.D. Originally, athletes only competed in one event, a sprint across the stadium. Eventually it grew to include boxing, wrestling, discus throwing, chariot races, and other events. The Olympics were resurrected in the 19th century with the establishment of the International Olympic Committee in 1894. They have been split into winter and summer games alternating every two years and hosted by various cities around the world.

Fascinating Facts

We've learned a lot about the human body so far—the parts of the cell, composition of bones, movement of muscles, and many other intricate design details. Let's give our brains a break and take in a few fun factoids. Check out these random facts that will amaze you, surprise you, or just make you say, "Hmmm."

The first heart cell starts to beat as early as three weeks after fertilization.

On average, your mouth produces enough saliva to fill a two-liter soda bottle every day!

Your ears and nose are always growing!

There are more bacteria living in your mouth than people living in the entire world!

The small intestine is somewhere between 18 and 23 feet in length. If it weren't packed tightly and looping back and forth, it wouldn't fit in your abdominal cavity.

It's more tiring to stand than it is to walk because when you stand you must use the same muscles for a longer period of time.

Humans shed about 600,000 skin particles every hour.

Every time you gain a pound of fat, your body adds seven miles of blood vessels to service the new tissue. If you lose the pound, your body will break down the unnecessary blood vessels and reabsorb them.

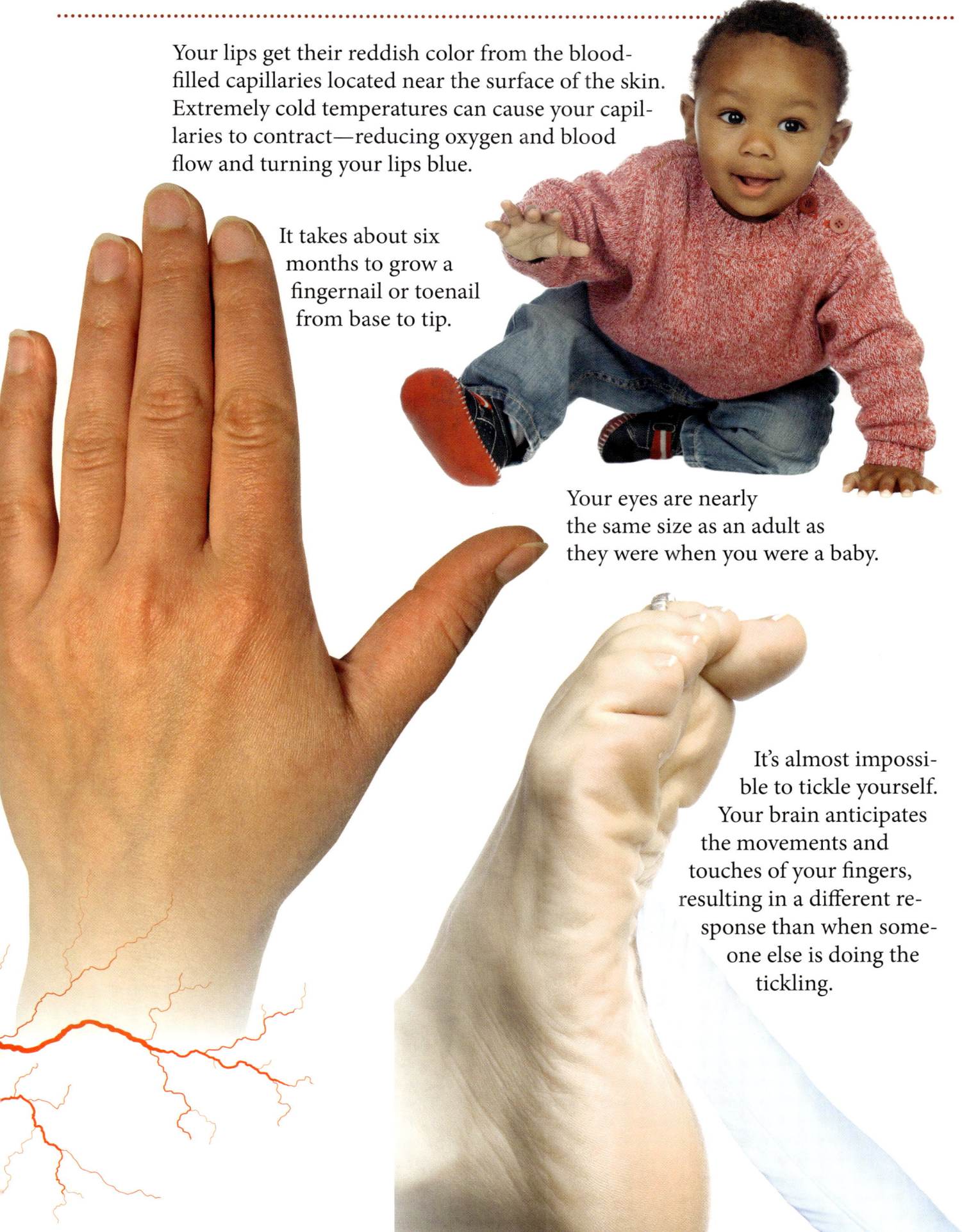

Your lips get their reddish color from the blood-filled capillaries located near the surface of the skin. Extremely cold temperatures can cause your capillaries to contract—reducing oxygen and blood flow and turning your lips blue.

It takes about six months to grow a fingernail or toenail from base to tip.

Your eyes are nearly the same size as an adult as they were when you were a baby.

It's almost impossible to tickle yourself. Your brain anticipates the movements and touches of your fingers, resulting in a different response than when someone else is doing the tickling.

Human Body, Tall and Small

The cause of variations in human height is still a scientific mystery. It is thought that 80% of a person's height is determined by genetics, while the rest is due to nutrition, lifestyle, climate, and other factors. Though we don't know everything about height, we can still study its effects and trends across different times and cultures.

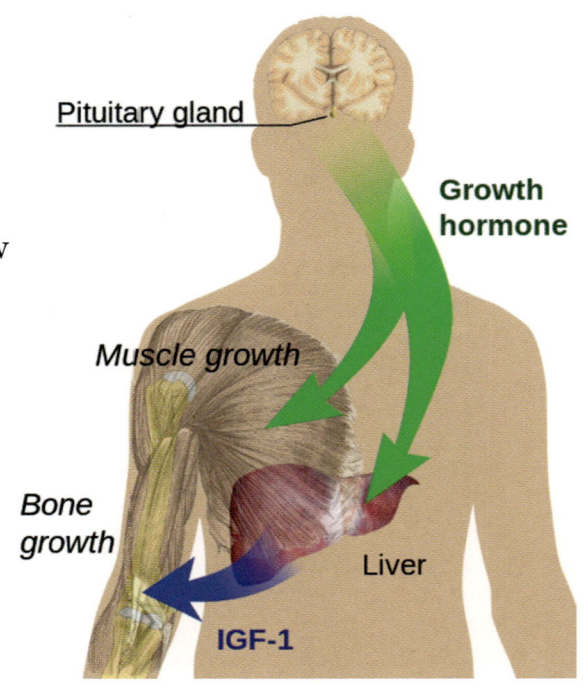

Pituitary gland

Growth hormone

Muscle growth

Bone growth

Liver

IGF-1

TALLEST AND SHORTEST

The tallest average population in the world are males from the Netherlands, whose average height is 6 feet. Females from Bolivia average 4 feet 8 inches, making them the shortest.

HEIGHT AND CULTURE

Comprehensive studies seem to show a correlation between a nation's average height and its economic prosperity, level of education, ease of access to nutrition, income, and so on. With the rise of developed nations over the past two centuries, the global average height of humans has increased.

The average male height is around 5 feet 9 inches. The average female height is around 5 feet 3 inches.

5 ft. 9 in.

5 ft. 3 in.

DWARFISM

The advocacy group Little People of America defines dwarfism as a medical or genetic condition resulting in an adult height of 4 feet 10 inches or shorter. Disproportionate dwarfism is defined as having an average-size torso with disproportionately sized limbs, or average-size limbs with a disproportionately sized torso. Proportionate dwarfism is defined as having proportionate torso and limbs—simply shorter than average.

GIGANTISM

Gigantism happens when a child's pituitary gland malfunctions due to a tumor. The pituitary gland starts pumping out an abnormally high amount of growth hormones, which causes the enlargement of hands, feet, toes, fingers, jaw, and forehead—and, of course, it makes the person very tall. Additional symptoms of the tumor are headaches and vision problems. Many people who suffer from gigantism do not have peripheral vision.

GOLIATH

The Bible says Goliath was six cubits and a span (1 Samuel 17:4). This would mean he was around nine to ten feet tall. Some have suggested that Goliath had something akin to gigantism or that his tall stature was due to demonic influence. We may not have access to a full understanding of how Goliath grew that tall, but there is no biomechanical reason why humans could not have grown to exactly the dimensions given in Scripture.

Anna Haining Bates (1846-1888) grew to be 7' 11½". Her parents (shown here) were of average height.

Serious Answers to Funny Questions

Studying the human body can provoke a ton of interesting questions. Are you hungry for answers? Well hold on to your hat, because scientists have already solved some of these puzzling mysteries.

WHY DO WE GET BRAIN FREEZE?

When you gobble cold ice cream or race your friend to finish a slushee, your head may start to hurt—a condition known as brain freeze! Nerves in your mouth take note of the extreme cold, telling blood vessels in the head to swell for warmth. Next time you get this temporary but painful headache, try drinking some warm water or sticking your tongue to the roof of your mouth to warm the cold-sensing nerves.

WHY DO WE BURP?

Everyone swallows a little air when they eat or drink. The stomach has sensors that detect when it is starting to get too full. To release the unneeded air, the esophagus relaxes a little, allowing the air to escape in a sometimes noisy burp! Drinking carbonated drinks and eating too quickly can cause you to burp more often.

DID YOU KNOW?

In some countries, burping after a meal is considered good manners. It tells the cook that you enjoyed the food.

WHY DO WE SNEEZE?

Achoo! A sneeze is born when sensors in your nose tell your brain that something is inside that needs to come out. An area located in the lower brain stem sends signals to the body to trigger a sneeze. The reflex starts by closing your throat, eyes, and mouth, squeezing your chest muscles, and then releasing it all quickly to force air out of your mouth and nose. With a speed of about 100 miles per hour, it still may take two or three sneezes before your nose can be cleared of the dust, bacteria, or debris.

WHAT IS THAT "HANGY-BALL" IN THE BACK OF MY THROAT?

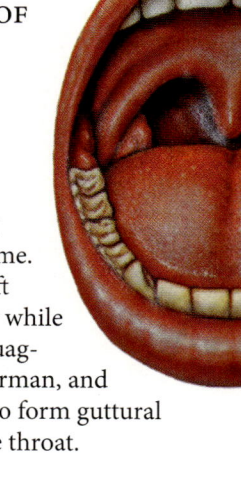

Its proper name is the palatine uvula. Though the uvula may serve more purposes than we know, it has been found to lubricate the throat by excreting large amounts of saliva at one time. It also works closely with the soft palate to seal off the nasal cavity while you eat and drink. Certain languages—such as Hebrew, French, German, and Arabic—also require the uvula to form guttural sounds involving the back of the throat.

HOW DO WE LAUGH?

When someone tells a joke, do you politely "tee hee," or do you slap your leg and let the "ha ha's" roar? Laughter is our body's natural response to humor. After analyzing the nature of a story or situation, the brain actually provokes the spastic movements, stifled giggles, and hearty guffaws that often follow something we think is funny. Laughter involves the contraction of multiple facial muscles as you smile, the opening and closing of the airway and voice box as you vocalize and gasp, and your tear ducts can even be activated if you're "laughing so hard you're crying."

"A merry heart does good, like medicine, but a broken spirit dries the bones." (Proverbs 17:22)

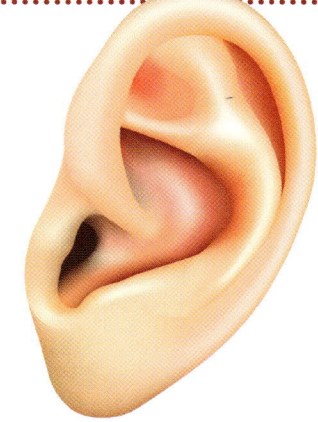

WHAT IS EARWAX AND WHY DO WE HAVE IT?

Earwax actually has a more official name: cerumen. It forms in special glands beneath the skin in the outer ear canal. It keeps your ears from drying out, kills bacteria, and traps dust, dirt, and other things to keep them from getting stuck deep within your ear.

WHAT IS HAIR MADE OF?

Hair is made of a protein called *keratin*. That's why you can get a haircut and not feel any pain. Each hair grows from a tunnel in the skin known as a *follicle*. The shape of the follicle determines how straight or curly your hair is. The part you can see is called the hair shaft and has three layers. The inner layer is the medulla, the next layer is the cortex, and the outside is covered in a layer of overlapping scales called the *cuticle*. The presence or absence of pigment cells in the cortex and medulla determine hair's color.

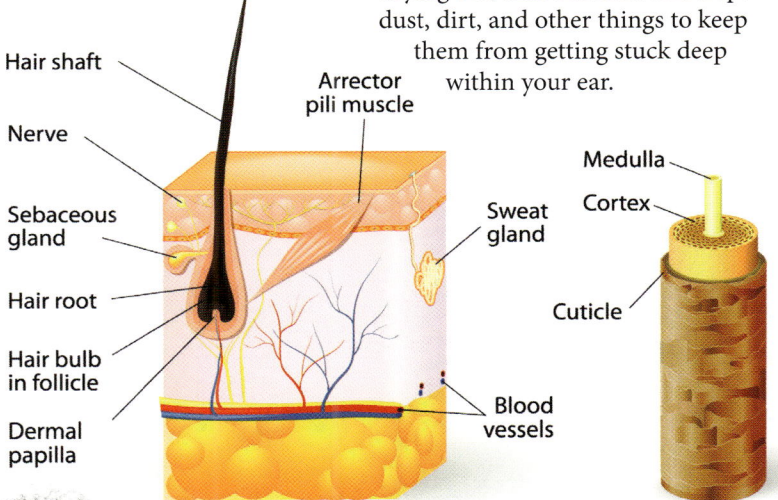

WHY DO WE HAVE EYEBROWS?

Whether they are fuzzy and thick or narrow and arched, eyebrows help shield our eyes from sweat, rain, and debris. They also cooperate with other parts of your face to form facial expressions, communicating your current emotions. Raised eyebrows may show surprise, while furrowed brows can display intense concentration, determination, or disapproval.

Out of This World

The human body is fascinating on its own—but have you ever wondered about how the human body functions in outer space?

THE HUMAN BODY IN SPACE

When astronauts go into orbit, they enter a weightless environment. Amazingly, their bodies have the ability to adjust to this environment, though it takes a few days. They must first overcome the initial shock from the lost functioning of their vestibular system, which provides a sense of balance. To compensate, astronauts must rely heavily on their vision to orient themselves in space, and that adjustment can take about six weeks.

Nutrition and exercise are important for everyone—regardless of where you're heading next! But launching into space demands that astronauts maintain a high level of physical fitness. They are monitored extensively with tests for certification before they launch. And as important as the preparation is, it's even more important that astronauts work to maintain the health of their bodies while in space because of the weightless environment.

If astronauts don't make up for the weight-bearing exercise they miss from daily life on Earth, their muscles will atrophy. They usually spend 45 minutes per day performing "resistive exercise," which includes lots of squats or dead lifts to strengthen the lower body. Another half hour is dedicated to running on a treadmill. A harness of bungees prevents them from floating away from the machine in the absence of gravity.

DID YOU KNOW?

NASA has a nutrition laboratory that prepares and develops all food for the astronauts to consume when they're in orbit.

NASA astronaut Col. Jeffrey Williams exercising on the International Space Station

DUPLICATING EARTH'S ATMOSPHERE

God designed very specific conditions in Earth's environment to support human life. Forests and seawater full of algae remove carbon dioxide and supply oxygen. Natural processes purify the atmosphere as well.

The International Space Station (ISS) must possess some of the same environmental conditions for the astronauts to survive there. For example, the space station has the same nitrogen-oxygen ratio as Earth's atmosphere, and carbon dioxide is regularly removed. Air contamination is monitored because trace contaminants might be toxic to the human body.

HUMAN BODY RESEARCH IN SPACE

Astronauts and cosmonauts conduct many experiments on the ISS, including the effect of weightlessness on the human body. As researchers identify negative effects, they seek ways to minimize them and develop countermeasures so astronauts can function in space for longer periods of time.

Sometimes, astronauts' visual acuity changes during their time in space. NASA provides several prescriptions of glasses to help them compensate. This eyesight problem could be related to fluid changes inside the eyeball that affect the optic nerve. Sodium levels at the cellular level could also contribute to the problem. We need further research to fully understand how and why outer space affects vision.

Optic nerve

NASA astronaut Col. Jeffrey Williams on the first time he returned from a long space flight:

"I could not close my eyes for the first hour or two because I…was interested in maintaining my balance. I could not stand up in darkness or with my eyes closed. I needed the visual cues to maintain my orientation and then use the muscles to maintain upright or I would fall over. So, it was very severe initially, but you improve very rapidly. You adapt certainly in the matter of hours."

RETURNING HOME

Returning to Earth significantly impacts the body in three primary areas.

1) Vestibular System—When astronauts return home, the gravity vector pulls the fluid in their inner ear, reengaging the vestibular system. But their brain hasn't been incorporating that balance sensor into its control system. It takes about two weeks for the brain to reintegrate the vestibular system into its regular function.

2) Muscles—Even though astronauts exercise every day, they can't work every muscle that they might use to move around on Earth. After landing, they have a protocol to work with trainers for 45 days on muscular shape, balance, and flexibility. It takes a month to six weeks for the muscles to recover 95% of their original condition after a stay of six months on the ISS.

3) Bones—Though astronauts can't feel a change in their bones, many suffer bone density loss that can take a year and a half to two years to recover.

Bone health is a very significant concern when astronauts consider the possibility of longer space flights—like going to Mars or to the moon for an extended period of time. Bone continually recycles itself, constantly breaking down and rebuilding at a balanced rate under gravitational conditions on Earth. When you go into a weightless environment, the rate of bone building stays essentially the same, but the rate of breaking down increases, causing the bones to atrophy. This process resembles osteoporosis.

Bone showing osteoporosis

The Uniqueness of Man

What separates man from the animal kingdom? Though we have some common traits that point to a common designer, there are significant differences that set human beings above the rest of creation. Our upright posture, lack of fur, scales or feathers, and ability to grip tools are accurate but superficial observations. Let's dive deeper to examine some traits that are unique to humans.

SPEECH, LANGUAGE, AND COMMUNICATION

No animal on Earth matches our capacity for expressing abstract ideas, and no animal's brain shifts into high gear for language-learning like that of a human infant. We are uniquely outfitted to use voices, bodies, and subtle facial gestures in seamless coordination to communicate.

APPRECIATION FOR BEAUTY

Humans display great appreciation for art, music, and beauty. No other creature shows this capacity. The flower is not impressed with its own majesty; it merely exists with no conscious awareness. Chimpanzees don't flock to see the enigmatic smile of the *Mona Lisa*, nor do the stars muse on the heavens they themselves grace. The fact that mankind alone is able to recognize and marvel at the world around him is a wonder in itself!

Research shows that when a listener deeply enjoys music, his brain employs the same chemistry that brings pleasurable sensations from certain other activities. Dogs, for instance, hear the sounds of music but do not recognize them as music and do not derive a similar pleasure from listening to them. This solely human characteristic, which has no apparent evolutionary purpose, makes sense if people were originally intended to enjoy their Creator through generating and enjoying pleasurable activities ordained by Him.

HUMAN-CHIMP COMPARISON

Evolutionists claim human DNA is 98-99% similar to chimpanzee DNA, but this information is very misleading. When researchers only compare parts of DNA already known to be similar in chimpanzees and humans, they get an impression of great similarity. However, when you compare all regions of the two genomes, the similarity is closer to about 81%.

In addition, the complex functional aspects of genes and their regulatory networks differ significantly between humans and chimps. These differences are now known to be a more significant factor in determining various traits than the gene sequence alone. Similar human and ape genes are expressed in very different ways and times to build our very different bodies.

MADE IN GOD'S IMAGE

Genesis 1 reveals that mankind was created in the image of God, a quality that separates him from the animals created on Day 6. This special creation explains why human behavior is far more complex than that of any other living thing on the planet. We reveal God's image in many ways. For example:

- We have an everlasting spirit capable of understanding morals and having fellowship with God.

- We can imagine and create objects never seen before.

- We can show compassion for strangers or enemies.

- We can ponder our role and fate in creation.

- We can know God.

God has gifted these abilities to mankind alone, His special creation, allowing us the privilege of displaying aspects of His nature. In fact, God treasures mankind so much that He sent the Lord Jesus Christ into the world to live as one of us. He died a torturous and brutal death to pay for our sins and rose again in order to reconcile us to Himself. It is this value that God places on man that truly separates us from the rest of creation.

We don't function purely by instinct like animals do. We have a will with which to choose whether to respond in love and worship to God or to reject Him as our Creator and Savior forever. With the life you have been given, what will you choose?

Humans and Chimps: Common Descent or Common Design?

When Charles Darwin published *On the Origin of Species* in 1859, he presented his case for the evolution of every living thing from a universal common ancestor—except for humans. A few years later, Thomas Huxley filled in this gap and wrote a book claiming that humans evolved from an ape-like ancestor. Since then, many scientists have assumed human evolution to be true. Part of their reasoning comes from the seeming similarities between humans and chimpanzees, thought to be close evolutionary "relatives." How accurate are these claims?

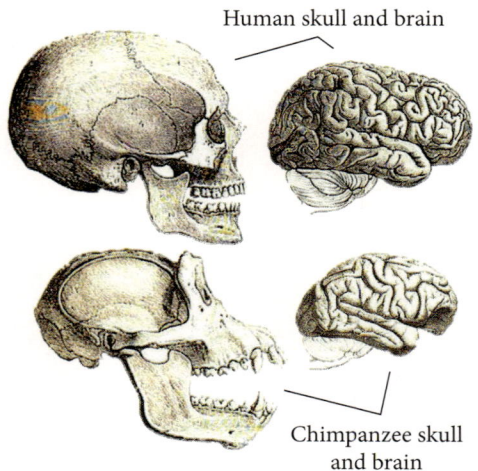

Human skull and brain

Chimpanzee skull and brain

GENOMES

Chimps and humans are often reported to be about 98-99% similar. However, this number is based on the comparison of DNA regions that are already known to be similar. DNA that is very different is typically ignored. When the entire genomes are compared, including the regions that are very different, a more realistic estimate is that they are only about 81% similar. This amount of difference is too great for evolution to account for in the three to six million years of change that supposedly occurred since humans and chimps allegedly split from a common ancestor.

BRAINS

Humans have a much larger motor sensory portion of the brain than chimps, meaning that we not only have specific muscles and hand features that allow precise and uniquely human movement, but we also have the mental motor capacity to control them. In other words, our "software" perfectly integrates with our particular "hardware." Not only that, we also have the mental ability to reason, create new ideas, and use complex language at a level that no chimp or any other animal can.

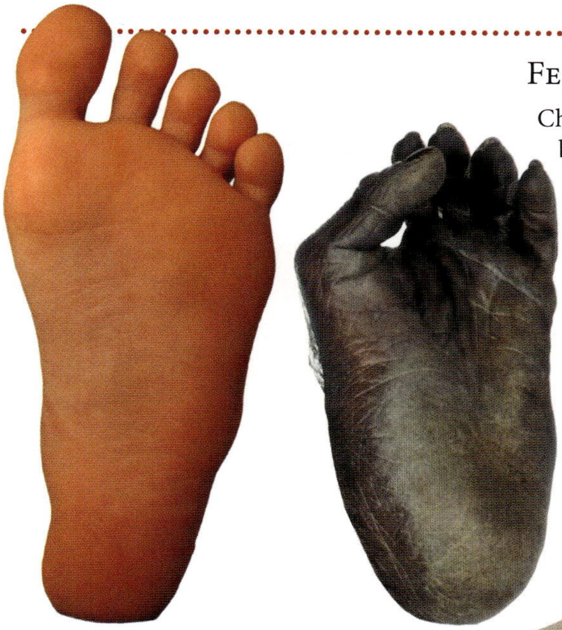

Feet

Chimps can rear back and walk on two feet for a short time, making them bipedal, but humans are habitually bipedal. It's perhaps as awkward for a chimp to walk on two feet as it is for a human to walk on all fours. Part of the reason is that human feet have a long arch from heel to toe that acts as a spring for walking and running. Chimps have flat, hand-like feet that are great for grasping tree branches but don't work well for walking long distances. Human knees face forward, perfect for walking, but chimp feet point outward.

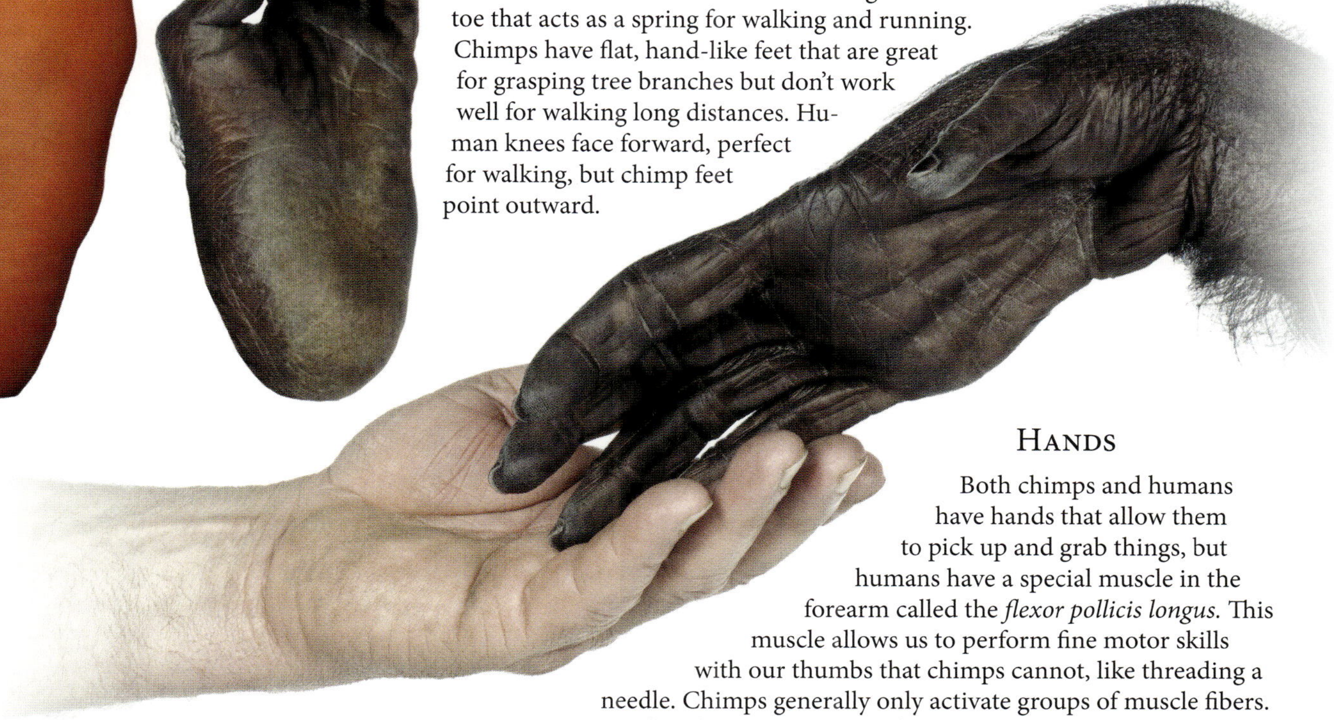

Hands

Both chimps and humans have hands that allow them to pick up and grab things, but humans have a special muscle in the forearm called the *flexor pollicis longus*. This muscle allows us to perform fine motor skills with our thumbs that chimps cannot, like threading a needle. Chimps generally only activate groups of muscle fibers. It makes their grip stronger than a human's but far less nimble.

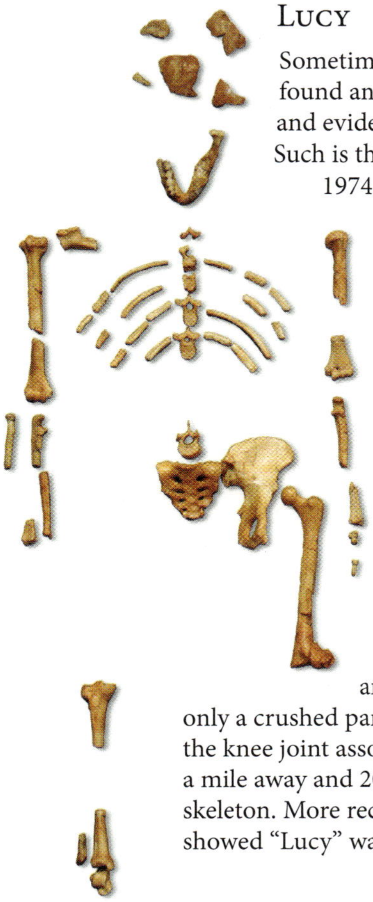

Lucy

Sometimes part of an ape skeleton is found and claimed to be part-human and evidence of man's evolution. Such is the case with Lucy. In 1974, a partial skeleton was found in Ethiopia. Hailed as a transitional form, it allegedly closed the gap between man and his ape-like ancestors. Since then, new evidence has challenged that claim. Most of the bones, including its skull and fingers, are more ape-like than human. Also, the ability to walk upright is based on the arrangement of the pelvis and knee joints. However, only a crushed partial pelvis was found, and the knee joint associated with Lucy was found a mile away and 200 feet deeper than the skeleton. More recent and more complete finds showed "Lucy" was merely an extinct ape.

Semicircular Canals

The semicircular canals tell a creature which side is up. In humans, they are perpendicular to the ground since we walk upright. In chimps and all other apes, they are positioned differently because of their skulls' orientation to their spines.

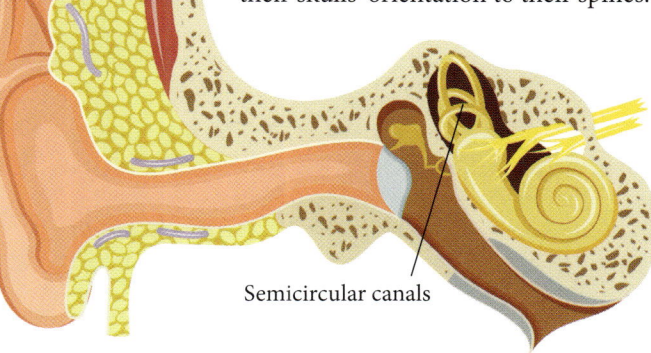

Semicircular canals

Common Design

While we have addressed numerous differences between humans and chimps, there are certainly some observable similarities in design as well—two eyes, a nose, a mouth, ears located on either side of the head, and so on. We could also point to various similarities in design between many creatures on Earth. Though evolutionists argue this as proof of descent from an ancient common ancestor, common design patterns fit perfectly with explanations that they originated from a common designer to serve similar purposes.

Bioethics

Whenever science advances in its understanding and examination of the physical world, moral questions come to the forefront. How far should we manipulate God's designed systems for the sake of research? What values should guide our decisions?

> "For you formed my inward parts; You covered me in my mother's womb." (Psalm 139:13)

The Bible teaches that man is a special creation (Genesis 1:27), life begins at conception (Jeremiah 1:5; Galatians 1:15), and the pre-born are human beings (Luke 1:41; Exodus 21:22-25). Numerous passages teach that since human life bears the image of God, it should be valued and protected. This stance on the value of human life at all stages helps us answer questions on bioethics as technology continues to advance. Humans, healthy or unhealthy, young or old, born or unborn, deserve protection. According to Scripture, the cessation of life must be called death, and willful death—even in the name of research—constitutes murder. Even when research does not involve the taking of life, we must still be careful to regard humanity, made in God's image, as valuable and respect the boundaries God has built into creation.

STEM CELLS

When a sperm and an egg combine, they form a zygote—the first cell of a brand-new human being. Zygote cells multiply enough to form a hollow sphere of cells called a *blastocyst* with an inner cell mass. The sphere will combine with cells of the uterus and form extra-embryonic tissues. The inner cell mass will begin developing into a recognizable baby.

Sometimes a zygote is artificially created and allowed to form into a blastocyst. Then the blastocyst is destroyed and the inner cell mass is removed. These cells are called *embryonic stem cells* (ESCs). They are valued very highly in scientific research because they can mature into any one of hundreds of cell types. But harvesting stem cells from a zygote will kill that person.

However, stem cells can also be taken from adults, called *adult stem cells* (aSCs). These do not require ending a human life, but they are able to turn into far fewer different tissues. They are more difficult to isolate and, unlike ESCs, cannot be cultured in a dish virtually indefinitely.

There are essentially no moral objections to using aSCs since they do not end a human life. In fact, dozens of helpful treatments derive from aSCs. None yet come from ESCs, which have been too difficult to coax into tissues. So why not use cells that do not harm a life *and* actually work?

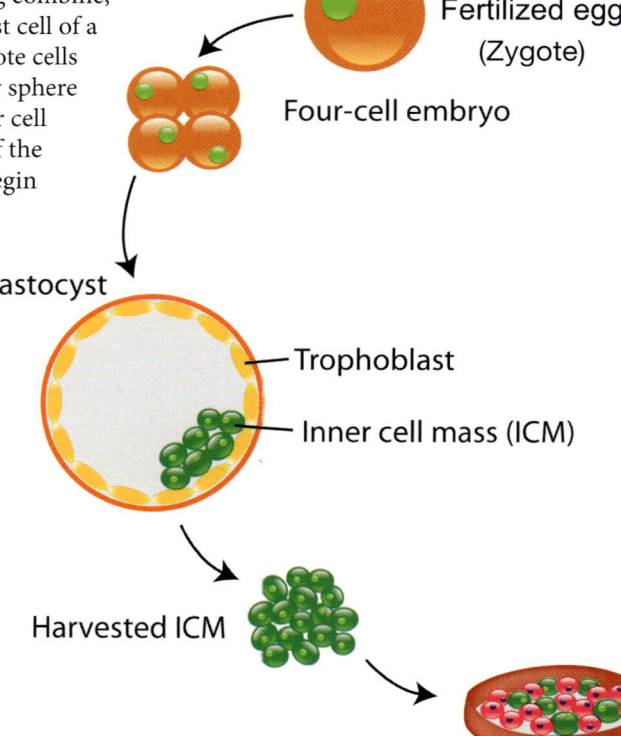

Fertilized egg (Zygote)

Four-cell embryo

Blastocyst

Trophoblast

Inner cell mass (ICM)

Harvested ICM

Embryonic stem cells

ANIMAL CLONING

Whole-animal cloning involves taking the nucleus of an adult's cell and inserting it into a female's unfertilized egg cell that has had its own nucleus removed. The egg with the adult nucleus is then inserted into a female's womb and allowed to develop. The result is a clone of the adult whose DNA is in the nucleus.

Cloning circumvents God's designed mechanisms, leading to animals that can become malformed or perish. This is not a problem morally or legally when animals are concerned but does become a problem when the clone is human. Human cloning would lead to all kinds of artificial and unfavorable family units and a high risk of a malformed baby. It also opens the door to simply developing babies for the purpose of harvesting body parts, an abhorrent abuse of precious human life.

DID YOU KNOW?

Dolly the sheep was the first animal to be cloned from an adult sheep cell. Born in 1996 at the Roslin Institute in Scotland, she lived only 6.5 years before she was euthanized due to unresolvable health issues.

ANIMAL/HUMAN HYBRIDS

Hybrids between animals and humans begin in a similar way to human cloning and stem cells. Researchers take a female animal's egg and replace its nucleus with that of a human adult cell—or vice versa. The supposed intent of this is to cross reproduction barriers that would otherwise not be possible via normal systems and also to obtain embryonic stem cells for medical research. There are currently animals in existence with partially human organs.

ABORTION

The term "abortion" as it is used today refers to the chemical or surgical termination of a human life before birth. Some abortion procedures are performed for medical reasons, but around 93% are done for non-medical reasons, such as being unable or unwilling to care for the child. Various techniques are used to abort the baby according to the level of development. Some abortionists use pills to destroy the developing baby, while others use forms of surgery and physical extraction during later stages of development. Once again, we look to the Bible—God's Word—as the final authority and see that any form of voluntary abortion is immoral because God cherishes life.

The Body of Jesus

The incarnation of Jesus Christ is a central miracle of the Bible. When He was born on this earth, Jesus was both true God and true man. God came to save us by becoming one of us. He became a real human being and walked and talked with us, yet He also remained God.

"'Behold, the virgin shall be with child, and bear a Son, and they shall call His name Immanuel,' which is translated, 'God with us.'" (Matthew 1:23; see Isaiah 7:14)

He understands what it's like to be human because He had a physical body like ours. Jesus was born and grew up just like you, but His Father was God and His mother was a young woman named Mary. Mary was a virgin, showing that Jesus' conception was miraculous. Mary descended from King David, therefore Jesus descended not only from Adam but also from the tribe of Judah.

"And the Word became flesh and dwelt among us, and we beheld His glory, the glory as of the only begotten of the Father, full of grace and truth." (John 1:14)

Being fully human, Jesus needed food and water like we do. At the beginning of His ministry, right after His baptism, the Spirit led Jesus into the wilderness. There all alone, He fasted and prayed for 40 days. Afterward, He was famished. Forty days is around the maximum amount of time a person can go without food.

Jesus knew hunger and thirst, and He knew loneliness and heartache. At the death of His friend Lazarus, He cried because He loved him so much.

"Then Jesus was led up by the Spirit into the wilderness to be tempted by the devil. And when He had fasted forty days and forty nights, afterward He was hungry." (Matthew 4:1-2)

"And He said, 'Where have you laid him?' They said to Him, 'Lord, come and see.' Jesus wept. Then the Jews said, 'See how He loved him!'" (John 11:34-36)

"Looking unto Jesus, the author and finisher of our faith, who for the joy that was set before Him endured the cross, despising the shame, and has sat down at the right hand of the throne of God." (Hebrews 12:2-3)

"But the angel answered and said to the women, 'Do not be afraid, for I know that you seek Jesus who was crucified. He is not here; for He is risen, as He said.'" (Matthew 28:5-6)

"But He was wounded for our transgressions, He was bruised for our iniquities…and by His stripes we are healed. All we like sheep have gone astray…and the LORD has laid on Him the iniquity of us all." (Isaiah 53:5-6)

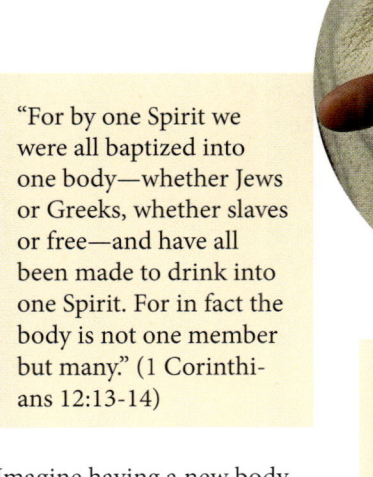

"For by one Spirit we were all baptized into one body—whether Jews or Greeks, whether slaves or free—and have all been made to drink into one Spirit. For in fact the body is not one member but many." (1 Corinthians 12:13-14)

Imagine having a new body that will never grow old. Our heavenly bodies will be like Jesus' resurrected and glorified body. Those who follow Jesus can look forward to an eternity in heaven with God and His restored creation!

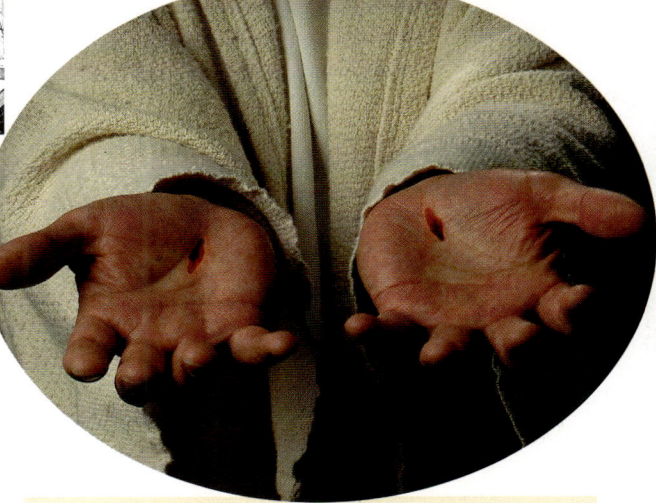

"'Behold, the tabernacle of God is with men, and He will dwell with them, and they shall be His people. God Himself will be with them and be their God. And God will wipe away every tear from their eyes; there shall be no more death, nor sorrow, nor crying. There shall be no more pain, for the former things have passed away.'"

"Then He who sat on the throne said, 'Behold, I make all things new.'" (Revelation 21:3-5)

Made in His Image

While animals may display God's creativity and power, only human beings are honored to bear His image. Some aspects of this truth are mysterious, but some ways we are made in God's image can be seen through Scripture.

> "So God created man in His own image; in the image of God He created him; male and female He created them." (Genesis 1:27)

CREATED WITH A SPIRIT

We inherit physical characteristics from our parents, but our spiritual attributes come from God. A person's spirit is immaterial, meaning it is not physical. The body and spirit function closely together until separated at death, when the spirit goes "to be absent from the body, and to be present with the Lord" (2 Corinthians 5:8). Since humans have the image of God, we can know and understand God (Jeremiah 9:24), connecting with Him through prayer and worship. We can also recognize and participate in moral virtues such as justice, mercy, forgiveness, grace, and truthfulness.

CREATED FOR WORK USING ABSTRACT THOUGHT AND CREATIVITY

God Himself worked for six days of the creation week. He is also at work today. Because work can be difficult, we often assume that it was introduced as part of the Curse. But God gave Adam tasks and responsibilities even before he sinned. In Genesis, God gave Adam the job of naming the animals and of tending and keeping the garden. After the Curse, accomplishing these tasks became a struggle, but that does not mean that we cannot still find great pleasure in our work. Unlike animals, we have the capacity to express creativity—art, music, science, industry, invention, athleticism, and more. We can use our imaginations to invent new technologies or design concepts. Completing tasks and projects using abstract thought and creativity reflects our Creator, whose grand works are displayed in everything we see.

> "Three times God insists that man was *created*. Man is not some higher order of ape. Man bears the image of the Creator. You and I are unique in all of creation. We are not the product of ages of random atomic interplay. The omnipotent and omniscient Creator personally designed us."
>
> —Dr. Henry M. Morris III,
> CEO of the Institute for Creation Research

CREATED TO RULE

God rules over the entire universe, but He chose to share His reign with man, granting him the authority over other creatures and responsibility for stewardship of the earth (Genesis 1:28). We see mankind exercising rule on the earth when people divert river water for irrigation, harness wind power for sailing and windmills, use rocks for buildings, herd livestock, ride horses, protect nature preserves for conservation, use timber for wood, and farm crops for food. Human authority is an extension of God's ultimate authority over everything.

CREATED TO EXPRESS MORAL-BASED EMOTIONS

God is not void of feeling. He has personality and is personal. He loves mankind, hates evil, takes compassion on the orphan, grieves over sin, is jealous for our souls and affections, and rejoices in truth and victory. He created us with the ability to experience these types of emotions as well. While we can be ruled by sinful or misplaced emotions, God's emotional responses are always right and fitting.

"And the LORD God said, 'It is not good that man should be alone.'" (Genesis 2:18)

CREATED FOR COMMUNION WITH GOD

God experiences perfect unity and communion within the Trinity—the Father, the Son, and the Holy Spirit. He also desires relationship with His most treasured creation, human beings. Likewise, humans desire a connection with God and other humans. We reflect God's relational nature when we commune with Him, or when a man and woman marry, or when we relate with each other in families, churches, and communities.

SIN, SALVATION, AND THE IMAGE OF GOD

The Bible indicates that even in our fallen state, all humans still bear the image of God. However, sin has significantly blurred our reflection of Him. We fall short of properly expressing His image in every area of life.

"For all have sinned and fall short of the glory of God." (Romans 3:23)

But the Lord Jesus Christ was crucified so that our spiritual nature that was "dead in trespasses and sin" (Ephesians 2:1) is renewed and we can truly live in ways that reveal the image of God. Those who turn from sin and put their trust in Him are given the Holy Spirit, who enables us to reflect God's image more fully. One day, we will be given new and glorified bodies like Jesus had after His resurrection (Philippians 3:21). On that day, we will be made in His image like never before.

"Beloved, now we are children of God; and it has not yet been revealed what we shall be, but we know that when He is revealed, we shall be like Him, for we shall see Him as He is." (1 John 3:2)

Using Our Bodies for God's Glory

When we turn from sin and put our trust in the Lord Jesus Christ, our salvation story does not stop there. In fact, it is only the beginning. God has called us to Himself to reflect His image and bring Him glory! As Ephesians 2:10 tells us, God has prepared good works for us to do as His new creation. Though this truth is deeply spiritual, carrying it out can be very physical. Any good works we do will require us to make use of the incredible bodies God has given us. What are some ways we can use our bodies for His glory?

> "For we are His workmanship, created in Christ Jesus for good works, which God prepared beforehand that we should walk in them." (Ephesians 2:10)

DISPLAYING DIVINE DESIGN

Did you know that God receives glory when you use your body as He designed it? Your performance as you swing a baseball bat, run a race, flip on a balance beam, or kick a ball showcases God's perfect design. When you apply your creativity and imagination in art or invention, it points to the existence of an even greater Inventor and Creator, the one who made you! And beyond physical or mental skill, you cannot overestimate the power of a smile on your face or a kind word on your lips to shine light in a dark world.

MEETING PHYSICAL CHALLENGES

Some people are hindered in their physical capabilities due to injury or illness. They have a unique opportunity to bring God glory as they maximize the use of their other abilities to compensate for the loss. We all face challenges and limitations in our lives and can demonstrate God's grace when we meet them with perseverance and faith.

Knowing and Doing His Will

God has equipped you with the very things you need to know and do His will. The coordination of your eyes and brain enable you to read and meditate on His Word. You can fill your mind with passages from the Bible that can help you navigate the many important decisions that confront you every day. And the brain's memorization capabilities are so vast you never have to worry about running out of memory! You can use your lips and voice and hands in worship and prayer to Him. Language and communication enable you to tell others of God's love and salvation.

A Living Sacrifice

"I beseech you therefore, brethren, by the mercies of God, that you present your bodies a living sacrifice, holy, acceptable to God, which is your reasonable service. And do not be conformed to this world, but be transformed by the renewing of your mind, that you may prove what is that good and acceptable and perfect will of God." (Romans 12:1-2)

Christlike behavior with our bodies will honor God. People will also notice if we do not conform to how the world encourages us to use our bodies. Presenting our bodies as a living sacrifice to the Lord Jesus is reasonable since He redeemed it for us by His death and He is Lord over us as our Creator.

"Let your light so shine before men, that they may see your good works and glorify your Father in heaven." (Matthew 5:16)

Serving Others

During Jesus' time on Earth, He modeled a life of sacrifice and service. Many of the good works God has prepared for us involve using our minds and bodies to help people, both inside and outside the church. Think about the gifts, talents, and abilities God has given you. How can you use them to minister to others? Strong arms can mow a lawn, carry groceries, or give an encouraging hug. Skilled hands can prepare a meal, build a house, or sew a blanket. Creative minds can imagine solutions to difficult problems. Faithful feet run to those in need, whether bringing food, clothing, or the good news of Christ. God has blessed us with an amazing body so we can be a blessing to others and point them to Him. Let's use everything—body, soul, and spirit—in a way that brings glory and honor to our awesome Creator.

Index

Image Credits

t-top; m-middle; b-bottom;
c-center; l-left; r-right

Bigstock: 8, 9ml, m, bm, br, 10, 11, 12l, br, 14b, tr, 15tl, b, 16-25, 26tr, 27, 28, 29tr, br, 30-37, 38bl, 39-41, 42l, tr, mr, br, 43-46, 47tr, m, 48-50, 51tl, br, 52bl, m, 53, 54bl, br, 55tr, b, 56, 57tl, bl, br, 58-59, 60tr, 61m, 62-67, 68t, 69t, 70b, 71, 72tr, b, 73, 74tl, bl, br, 75-76, 77mr, ml, b, 78bl, 79-85, 86tr, bl, 87bl, br, 88, 89tl, tr, 90br, 91-93, 94bl, br, 96, 97tl, tr, bl, 99t, r, 100, 101bl, br, 102b, 103tl, tm, tr, br, 104tl, 106l, 107br, 108-111

Fotolia: 52tr, 72ml, 90bl

Denise Franklin: 90t

Institute for Creation Research: 78m

iStock: 26l, br, 29l, 42bm, 54tr, 57tr, 68-69b, 78t, 97mr, 101tl, tr, 105, 107tl, m, bl

Ad Meskens: 12m

NASA: 9bl, 98, 99b

National Institutes of Health (NIH): 51tr

Public Domain: 9tm, tr, 12tr, 13, 15tm, tr, 47bl, 61tm, 77tm, 86br, 87tr, mr, 94tr, 95, 102tr, 103bl, 106r, 107tr

Travis Whitfill (Dallas Baptist University): 61r

Susan Windsor: 38r, 60br, 61tl, 70tr, 74tr, 89br, 104br

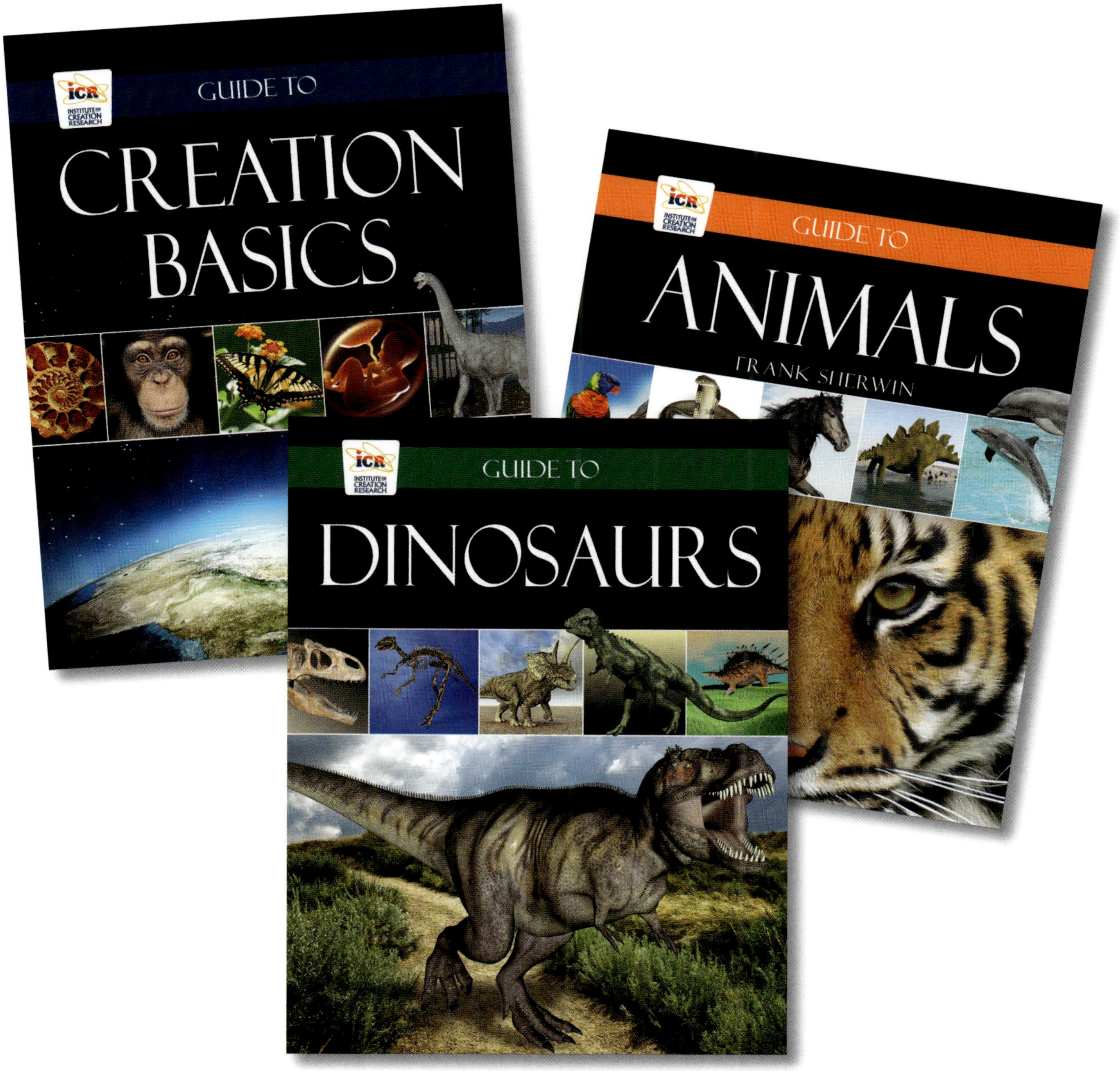

Get more facts with *Guide to Creation Basics, Guide to Animals,* and *Guide to Dinosaurs*!
Designed for all ages, these hardcover books are loaded with cutting-edge
scientific information and hundreds of full-color illustrations.

To order, call **800.628.7640** or visit **www.icr.org/store**

Also available for Kindle, Nook, and through the iBookstore

MADE IN HIS IMAGE

A four-episode DVD series on the complexities of the human body

This DVD series contains **English** closed captions and subtitles in **English, Spanish, Chinese, Arabic,** and **Korean**!

"God created man in His own image."
— Genesis 1:27 —

*M*ade in His Image, ICR's new DVD series, takes audiences on a journey through the most complex and miraculous creation on Earth—us! There is no better example of complex, conscious design than the human body. Featuring medical, engineering, and other experts, *Made in His Image* fascinates audiences with mind-blowing facts, dazzling imagery, and memorable illustrations.

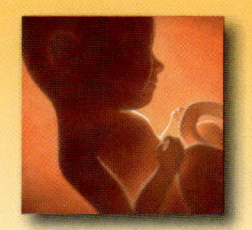

Episode 1: The Miracle of Birth. Only a masterful Creator could enable a baby to thrive in a watery world for nine months then suddenly live in an air-breathing environment at birth. Witness His incredible design from gestation to birth.

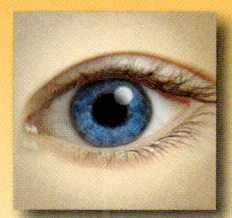

Episode 2: The Marvel of Eyes. The intricate engineering of the human visual system is vital for cognitive development from infancy through adulthood.

Episode 3: Uniquely Human Hands. Human hands and muscles display purposeful design, granting us unique abilities controlled by a sophisticated nervous system.

Episode 4: Beauty in Motion. This final episode illustrates the peak of human ability through athletic performance and highlights the aspects of complex design that confirm divine creation.

To order, call **800.628.7640** or visit **www.icr.org/store**